WATER

Exploring the Blue Planet

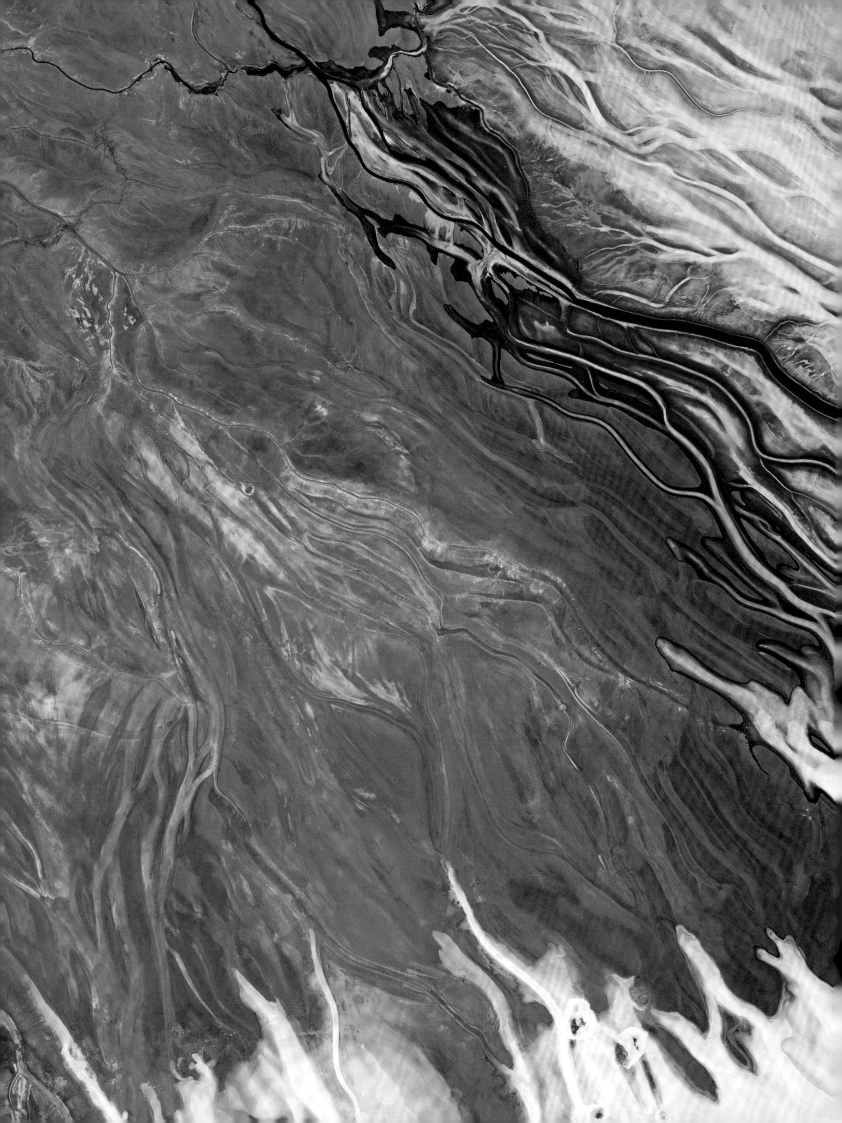

WATER

Exploring the Blue Planet

MARKUS EISL
GERALD MANSBERGER
PAUL SCHREILECHNER

FIREFLY BOOKS

A FIREFLY BOOK

Published by Firefly Books Ltd. 2016

First printing

Publisher Cataloging-in-Publication Data (U.S.)

Names: Eisl, Markus, author. | Mansbeger, Gerald, author. | Schreilechner, Paul, author. | Read, Beatrix, translator.
Title: Water : exploring the blue planet / authors, Markus Eisl, Gerald Mansberger, Paul Schreilechner ; translated by Beatrix Read.
Description: Richmond Hill, Ontario, Canada : Firefly Books, 2016. | Includes index | Originally published by Frederking & Thaler,
    Germany, as Wasser: Entdeckung des Blauen Planeten | Summary: This book discusses water and the current water predicament.
    It focuses on the health of our planet regarding water and uses technologically advanced systems to reveal the damage and
    destruction brought on by humans.
Identifiers: ISBN 978-1-77085-813-8 (hardcover)
Subjects: Water – Environmental aspects. | Water conservation.
Classification: LCC TD388.5E475 IDDC 333.9116 – dc23

Library and Archives Canada Cataloguing in Publication

Eisl, Markus, 1964-
[Wasser. English]
    Water : exploring the blue planet / Markus Eisl, Gerald
Mansberger, Paul Schreilechner ; translator, Beatrix Read.
Includes index.
Translation of: Wasser : Entdeckung des blauen Planeten.
ISBN 978-1-77085-813-8 (hardback)
    1. Water--Environmental aspects. 2. Water-supply.
3. Water conservation. I. Mansberger, Gerald, 1960-, author
II. Schreilechner, Paul, 1965-, author III. Read, Beatrix,
translator IV. Title. V. Title: Wasser. English.
GB661.2.E4813 2016   333.91   C2016-902992-1

Published in the United States by
Firefly Books (U.S.) Inc.
P.O. Box 1338, Ellicott Station
Buffalo, New York 14205

Published in Canada by
Firefly Books Ltd.
50 Staples Avenue, Unit 1
Richmond Hill, Ontario L4B 0A7

Translation: Beatrix Read

Printed in China

Responsible: Dorothea Sipilä
Editorial office: Albrecht Heinz, Munich
Idea and conceptual design: eoVision, Salzburg
Picture editing: Judith Grubinger-Preiner, eoVision, Salzburg
Layout concept and design of cover: coverdesign uhlig, Augsburg
Layout and composition: eoVision, Salzburg
Reproduction: Repro Ludwig, Zell am See
Production: Bettina Schippel
Overall production: Publishing house GeraNova Bruckmann

Front cover: Manhattan, USA
Back cover: Durrat Al Bahrain, Bahrain (top); Santorini, Greece (bottom)
Pg. 2/3: Current pattern at the edge of Salar de Uyuni, Bolivia
Pg. 4/5: Lighthouse Reef, coral reef in the Caribbean Sea, of the coast of Belize

CONTENTS

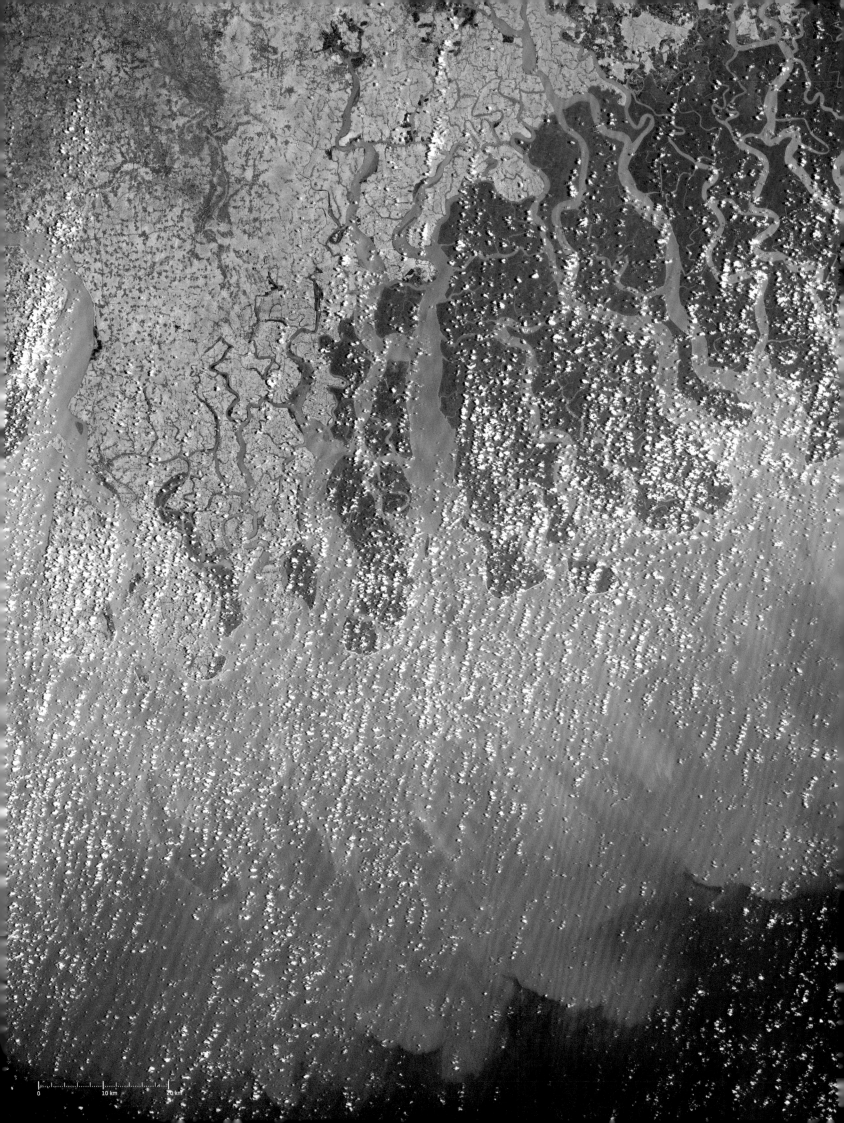

DIVING INTO A NEW WORLD

In the history of humans, the availability and use of water, as well as the mastery of its force, were always of central importance. This has not changed in our modern times, in which we have become accustomed to everything, or at least many things, being achievable and controllable. However, according to the United Nations, there are currently 700 million people without access to clean drinking water and it is expected that water consumption will increase by an additional 40 percent by 2030. At the same time, numerous people are exposed to the destructive impact of water again and again. In addition to the losses of human lives, the devastating results of floods and tsunamis include growing material losses. Due to the great importance of water, the United Nations proclaimed the "Decade of Water" (Water for Life) in 2005, which ended in 2015, and numerous initiatives around this topic were introduced.

Against this backdrop, it was only a small step for us to present the topic of water in an illustrated book, which exclusively shows satellite images. For space also has a lot to do with this topic; the images from space have formed our image of the blue planet, and it is also the data that has been acquired at very great heights that helps us now, in numerous application areas, to sensibly handle the resource "water." In addition to the colorful content of our presentation, which naturally must be restricted to a representative excerpt of the multiplicity of this wide topic, the aesthetics of the pictures has taken a center stage — and water lends itself also from this perspective, particularly well for an illustrated book.

In this book, we present in five sections: examples for the influence of water on landscapes, its role for settlements, its meaning as a basis of life for humans, the use of water for the infrastructure and water in connection with natural landscapes. Introductory descriptions of the subject areas and short descriptions of pictures complement the images of the Earth's reality provided by satellites. Views from experts, who comment on individual important topics in brief contributions and thus put the pictures in a greater context, complement each image.

At the beginning of the third millennium, we are at the crossroads in several respects — and in many of the relevant processes and decisions, water plays a role. We hope that we can contribute to new insights into the world of water with this illustrated book and thus expand the horizon concerning this important topic a little. We hope our readers make interesting discoveries when they view the fascinating satellite pictures.

Markus Eisl, Gerald Mansberger, Paul Schreilechner

← **The controlling element**
21°39'N · 88°37'E (April 26, 2001)

Cloud fields, ocean currents, rivers, swamps, rice fields — water dominates the satellite picture of the Ganges Delta at the Bay of Bengal in a multitude of ways. Water rich in sediments flows from numerous river arms; it partially originates from the Himalayas and mixes in fine shades of color with the waters of the Indian Ocean. In the areas of the delta, the dark green mangroves of the Sundurban National Park border on intensively farmed areas, which owe their fertility not least to the extensive floods during the monsoon season.

ABOUT THE IMPORTANCE OF WATER

Without water, nothing is possible for humans on Earth; humans can only survive a few days without this element. The importance of water for humans is visible everywhere — from agriculture to energy generation and from fishery to the protection of coasts. However, beyond the material domain, its importance also shows in cultural achievements — in works of world literature or the architecture of fountains and parks. Water plays a central role for us humans, but the same as all resources, it is not available in any desired volume. This makes it all the more important to handle it carefully.

The attractive effect of water, which by far exceeds its role as a vital substance, starts in childhood. You will hardly find a child that passes up the opportunity during play to deal with the diverse properties of this element and to explore its possibilities. And when people intend to realize their desires in recreation and vacation, these desires are reflected in the choice of their vacation location or recreational activities. When examining the most popular tourist destinations, as well as most favorite recreational activities (sports, for example), you will immediately notice that a large number of them have to do with water. Summer vacations with sun, beach and ocean, but also winter vacations (like skiing resorts in the Alps) would offer little attraction without water in its different manifestations.

Although water is a natural part of everyday life and in a way, a given, it still has a fascinating appeal at many levels. From microscopically small ice crystals that captivate you with an endlessly great variety of shapes despite the regularity of their hexagonal symmetry, and the roaring vehemence of large waterfalls to the elementary natural catastrophes of floods over large areas and tsunamis, which tens of thousands of people are helplessly exposed to, the fascination of water lies in its beauty, the same as in its terror.

From old stories

The fascination of water is also reflected in the great narratives of

> "Water is the principle of all things because water is everything and everything returns to water."
>
> Thales von Milet, approx. 625–547 BCE

← **Watercolor in the Gulf of Mexico**
29°10' N · 89°17' W (January 12, 2014)

The high sediment content after severe precipitation in the drainage basin of the Mississippi colors its water in brown. At the mouths, it mixes with the seawater and is distributed into fine color shades by the currents in the Gulf of Mexico. The side by side of nature and use by humans, as well as the incredibly variable appearance of the delta, make it a prime example for the multiplicity and changeability of water.

humans, in legends and myths, which conceptualized the beginning and end of the universe, and humans, in powerful stories and thereby assisted in prescientific times, to understand the cast in the great theatre of the world. Gods took the place of natural forces and were thus responsible for natural phenomena — regardless of whether it concerned lightning and thunder from Zeus, Jupiter, Thor or Taranis, or whether the gods of individual waters had caused the drowning of people who had approached the water in a careless or unrespectful manner.

Even today, many waters are associated with the names of these gods, including the oceans, the name of which originates from Okeanos, a titan from whom numerous river gods descended. In particular, the creation myths of many cultures refer to water that is at the beginning of all things, even before the actual creation of the world begins. By the same token, the end of life and end of the world is often associated with water. The events around Noah's Flood or the old Greek image of the rivers of the underworld are brought to mind. Water is not only the element of life, but also the element of death, and the amazing rainbow that appears as a symbol of hope is closely connected to the element water, too.

About the creative power of water

In mythology, water is often mentioned in connection with the creation of the world. And in the scientifically founded reality, water often plays a role in explaining the origin of forms and structures on Earth. In many cases, the creation and vanishing of landscapes is based on processes for which water provides an important, if not the most important, contribution; this includes weather and climate phenomena, the same as the effects of water in geophysical processes.

The defining importance of water for shaping landscapes is based on its physical and chemical properties, which allow it to develop an enormous dynamic force in combination with the energy from the sun and the gravity of the Earth, moon and sun. Rivers transport rocks to the sea that have broken off due to frost, the gravitation of the moon lets destructive flood waves surge at coasts and recurring ice ages result in mountains being smoothed down by ice crusts. The quick cycle of water with evaporation, precipitation and flowing back to the sea finds its mirror image in the cycle of mountain formations, erosion and the formation of sedimentary rock, which is lifted to mountains in geological periods.

The results of the design power of water are superbly visible in satellite images. From the spacious landscapes of valleys in Greenland, which are formed by outlet glaciers, to the fine ramifications of the drainage patterns in the silt areas of tidelands, the traces of water can be found over a wide-scale area.

Nature paradises at the water

Vacation paradises at the water are often equated with unspoiled nature. Colorful coral reefs with shoals of multicolored fish or clear mountain lakes in picturesque mountain landscapes are some of the most popular photogenic scenes and support the widespread illusion of an everlasting ideal. However, nature is anything but static.

Even before humans started to redesign their environment, the system "Earth" was in a dynamic balance: The environmental conditions usually changed slowly, thus allowing natural systems to adjust. This included catastrophic exceptions again and again. Examples are the Chicxulub meteorite, the impact of which (near the coast of Yucatán about 65 million years ago) is associated with the end of the dinosaurs, or the increase of cyanobacteria driving photosynthesis, which increased the oxygen level in the atmosphere 2.4 billion years ago, thus causing mass extinction, but creating conditions for the development of higher life forms at the same time.

Today, the world population is growing at a fast pace and the natural systems can no longer keep up with the enormous consumption of the resources soil, air and water. This makes it all the more important, especially in the context of water, to protect large enough areas from humans in order to create refuge for endangered species. Satellites may contribute by providing information about the condition and development of natural habitats.

Water and humans

Everybody knows about the invigorating effect of a glass of water on a hot summer day or after a long period of physical exertion. Humans can only survive a few days without water, a far shorter time period than without food. And it is important to humans that all other creatures are also dependent on water, since we live from vegetable and animal food. Only the cultivation of agricultural crop and livestock breeding allow the supply of a population in the current size, and this also requires the supply of plants and animals with nutrients and water. The large-scale reshaping of the Earth's surface into agricultural areas is one of the most noticeable marks humans have left on Earth. The irrigated fields catch one's eye when viewing an image from space — 70 percent of the water that

Blue spots in the sea of sand →
22°03' N · 54°38' E (October 28, 2010)

The largest sand desert of the world, Rub-al-Khali (which is located in Saudi Arabia for the most part), stretches over hundreds of miles with sand dunes that can reach up to 1,000 feet (300 m) in height. The saliferous clay of the sebkhas between the dunes that have been exposed by the wind, used to constitute the floor of a body of water; now, however, they are only covered in water occasionally.

is taken from the rivers, lakes and groundwater bodies of the Earth is used for agricultural use.

Another area where water satisfies human demands is fishery, which represents an expressive example for the finiteness of the Earth's resources at the same time. Since the reproduction rate of fish cannot keep up with the increasing fishing quota, fish stock is declining in many areas of the oceans. Aquacultures are considered the solution for this dead-end scenario and they dominate the satellite images of many coasts. Both sectors (agriculture and fishery and fish farming) are of central importance for the future of humans. The restrictions by nature must be considered all the more, so that food is ensured for humans for the future, too.

Water and technology

The properties of water make its control and use simple and difficult at the same time. Channels and hydropower plants are therefore seen alongside large structures that are intended to tame the destructive force of the water.

The large hydropower plants, like the ones at the Itaipú and Yangtze with their millions of cubic yards of concrete, are some of the most extensive infrastructure structures in the world. With capacities in the range of gigawatts, they are able to cover the electricity requirement of entire countries. However, the side effects of these projects are also extensive; they may reach from the relocation of entire cities and the destruction of ecosystems to the disturbance of the geological stabil-

ity. Similar effects, although to a far lesser extent, can also be noticed for the large traffic facilities at water. For example, harbor facilities can dominate large cities entirely with respect to their area. In many areas of the world, the facilities intended to protect the living space of humans from damages caused by water are even more extensive than the facilities for the utilization of water. One prime example is the levee systems in the Netherlands, which protect a large area of the country against floods, similar to the dams at the Mississippi that protect New Orleans.

Space technology should also be mentioned as a technical solution with respect to water. Over the past few decades, weather and Earth observation satellites have contributed to an incredible extent to the understanding of processes in the atmosphere and in the oceans, in which water plays a role. And it is also thanks to space activities that an overall view of our home planet is possible. This is impressively symbolized by the known image of the blue planet. Where does our journey lead us? At the beginning of the third millennium, humans are challenged in many areas with momentous developments and decisions that affect basically all areas of life due to their multifarious consequences. From climate change and the resource question — in particular relating to energy — to the further development of the economic systems and society in general, solutions must be found in the near future that enable the continued existence of humans in livable conditions in the long term. In many of these questions, water and our handling of it, plays a decisive role. We all bear the responsibility.

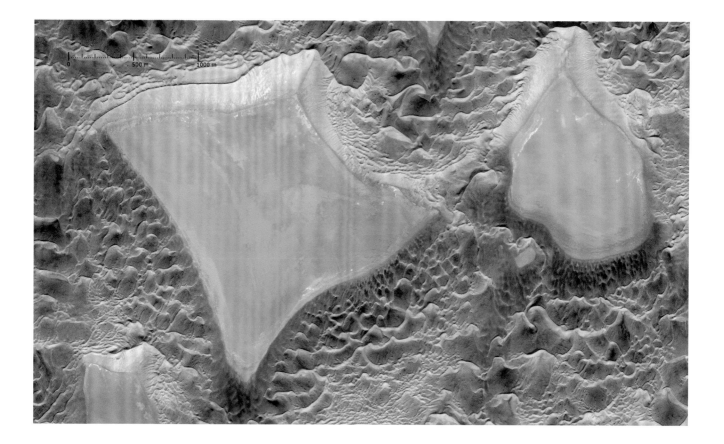

Water: The Architect

FORMED BY
WATER

MEANDER BENDS IN ROCK: Above Lake Powell, the Colorado River runs through the Canyonlands National Park. The sequence of sediment layers through which the river worked its way can be clearly tracked on the steep walls of the canyon (38°14' N · 109°49' W; April 22, 2013).

THE CREATIVE POWER OF WATER

It was a sensation in 2004, when the space probe *Mars Express* found proof that water exists around the south pole of the red planet in the form of ice in addition to frozen carbon dioxide. The importance of finding water on our neighbor planet has two reasons: The availability of usable water would considerably facilitate manned missions to Mars that take a long time, plus hopes to find traces of life on Mars received a boost.

The discovery of water on our neighboring planet successfully ended a long search, since spoors in Mars' landscape which were found a lot earlier, led to the assumption that water exists. The visible impact of the forces of water is far more noticeable on Earth than on Mars. Whether in the bizarrely ragged rock formations of the Bryce Canyon in Utah, the glacial valleys in the Alps or the long-stretching sand beaches of the North Sea, but also the dust-dry, sand-filled wadis of the Sahara — a glance at the Earth's surface shows the creating effect of water almost anywhere. The traces of its working can be found in large-size ranges, from the fine ramifications that the flowing surf leaves on a beach, to the wide-stretching river landscapes of the Amazon Basin. And yet water is only one, although very significant, component of the powers that formed the landscapes of the Earth's surface and will continue to do so.

Shifting surface

The strongest shifts of masses in the Earth's crust, which is only approximately 25 to 37 miles (40 to 60 km) below the surface of the continents, were caused by impacts from meteorites and asteroids in the early beginnings of the planet. Soon, those impacts became rarer and today one can only detect a few traces of such impacts in the landscape of the Earth's surface. As a result, streams in the glowing, viscous inside of Earth became the engine for the movement of the tectonic plates. Lifting, lowering and folds in the Earth's surface due to the plates moving against each other, formed the raw material from which the forces of erosion created a hardly comprehensible diversity of different landscape forms. In addition, volcanic activity is closely

"Nothing in this world is softer and weaker than water. Yet, to attack the hard and strong, nothing surpasses it." Laotse, Tao Te King

← Rock striations

56°23' N · 7980°24' W (April 24, 2001)

The approximately 1,500 islands of Belcher Islands appear like streaks of color in Canada's Hudson Bay. Softer layers disappeared from the surfaces of the sediment rock that was smoothed down by the glaciers from the Ice Age; now, the remaining hard rock layers form the striped structures of the islands. At the time when this picture was taken, at the beginning of May, the ice around the islands breaks and floats in floes across the bay.

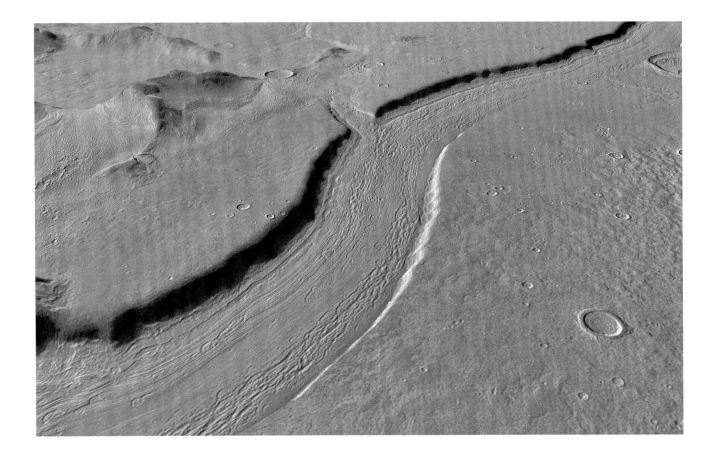

connected to the tectonic forces of the plate drift, which also leads to the creation of mountains. One of the most spectacular phenomena on our planet is the volcanic eruption — eruptions with an archaic force with which liquid and glowing rock is moved from the interior of the Earth to the surface. Craters of volcanoes, which often display a very symmetrical structure, also represent particular attractive landscape elements in satellite images.

Constant dripping wears the stone

Tectonic forces that resulted (and still result) in the unfolding of mountain ranges, such as the Alps and the Himalayas for example, are counteracted by balancing erosion forces that wear down the created mountains and fill hollows of valleys with the worn materials. In these processes, water plays a central role. Over decades to millions of years, even the hardest rock is cracked open by frost and chemically dissolved in water, smoothed down by glaciers and rivers, broken up into smaller and smaller parts and transported down toward the

valley over many miles until they are finally deposited somewhere. This contributes to things like the creation of a new valley floor or the expansion of a river delta at the coast of an ocean.

At the coasts of the bodies of water, surf, ocean currents and the constant turning of ebb and flow wear down the steep coasts and form wide sand beaches. Besides multivaried coast forms, the forces of erosion also create a wide range of the most diverse mountain and valley shapes, which are immediately noticeable on the satellite images in an impressive way.

Gone with the wind

Water, in its different forms of appearance, is not the only part of these processes. Winds and storms also grind down on rock due to the sand and dust particles they carry. They form sand dunes in the deserts of the world in a multitude of shapes — from individual crescentic dunes and dune chains that can reach over hundreds of miles, to the complex star dunes of the sand seas in the Sahara and in the

↑ **Water on the red planet**
Coordinates for Mars: 41° S · 107° E (May 14, 2012)

This diagonal view was generated from data from the Mars probe *Mars Express*. It shows Reull Vallis, a valley that is half a mile to 4 miles (1–7 km) wide and 1,000 feet (300 m), deep and that was presumably created as a water-bearing river valley. Long structures in the valley are interpreted as traces from glaciers in a later development period of the red planet.

Sunken glacier valleys →
61°25' N · 5°34' E (July 21, 2000)

The deep cuts of the fjords give the west coast of Norway its unmistakable character. The valleys, which are now flooded by the sea, were formed by glaciers. They often stretch over 60 miles (100 km) to the interior of the country and some are framed by rock walls that are considered some of the highest cliffs in the world.

Rub-al-Khali of Saudi Arabia. In addition, storms often carry fine sand and dust particles over thousands of miles into areas far away from their location of origin. One important example is dust transports from the Sahara, which reach across the Atlantic to the South American Amazon Basin by means of far-reaching storm systems, where they play a significant role in the nutrient inflow of the rain forest.

Life shaping the environment

Another important factor and, as far as we know, only existing on Earth, for the development of landscapes, is life, which as a component of its surroundings, affects them. Effects from vegetation on the surface of the Earth include, besides erosion-increasing effects, also those that significantly contribute to the stabilization of the ground. On the one hand, plants with their roots are able to expand existing cracks in rock and thus noticeably accelerate the destruction of rock. On the other hand, however, they play an important role in fixing the soil and stony subsoil at its location and thus counteract the wearing effects of water after heavy downpours.

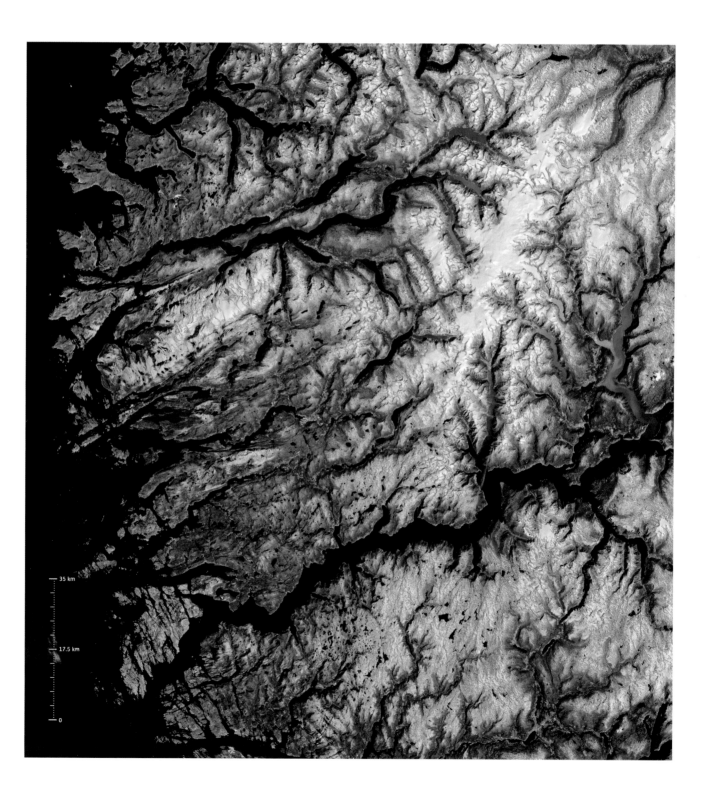

35 km

17.5 km

0

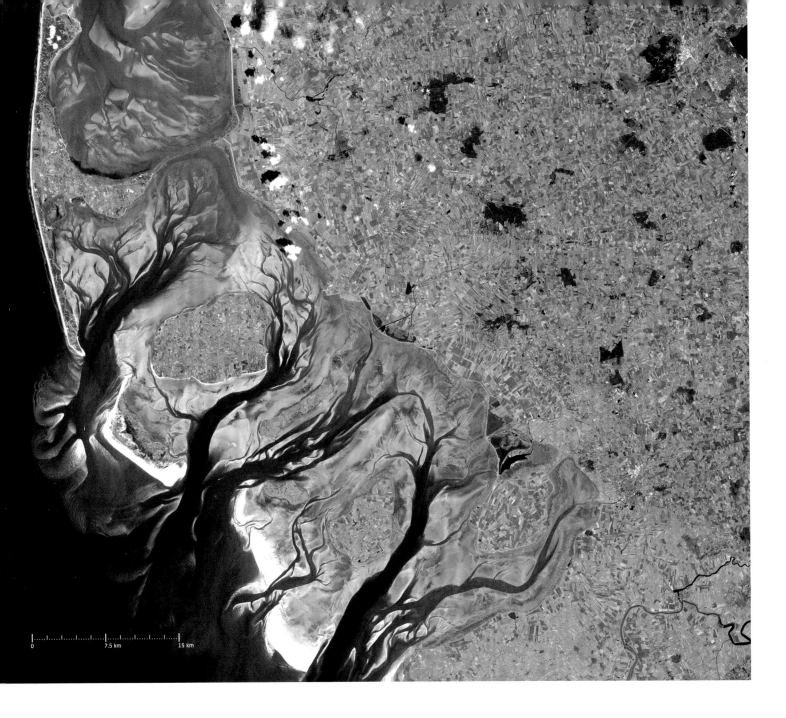

0   7.5 km   15 km

One example to show how landscapes can develop after losing the protective vegetation layer can be found in the karst landscape along the Adriatic Sea, where the demand for wood for shipbuilding resulted in the clearing of the original forests during the Roman Empire. Rain quickly washed away the humus layer that was no longer kept in place by roots from the water-permeable rocks of the limestone mountains, and as a result, the limestone was directly exposed to erosion. Since the forming of a humus layer is a lengthy process, it is very difficult to reverse such degradation of the ground.

Another example of the landscape-forming power of plants is their role in creating wetlands and swamps, which fill and replace lakes bit by bit. Plants are not the only living organisms that contribute to forming landscapes. The mentioned limestone rocks would not exist without myriads of fossilized calcareous skeletons from mussels, corals and other sea dwellers that lived many millions of years ago. Today, we also experience how reefs grow due to the constant effect of corals. Surely, the best known example is the Great Barrier Reef, which stretches over a length of 1,430 miles (2,300 km) in front of the coast of northeast Australia. With their numerous small islands, these coral reefs represent important landscape elements and habitats for a multitude of animal species, and at the same time, they protect the coasts behind them when storms stir up the surf to destructive forces. One special form of these reefs are the ring-shaped and often very small atolls, some of which are hardly higher than sea level; the vulnerability of which is also apparent in satellite images. These habitats are notably endangered by an increase of the sea level and water temperature. Larger animals are also involved in shaping landscapes. The beaver, a native animal of Canada and Europe, is known for being able to change entire regions because of its building activities in water.

Beavers block small rivers with extensive dams and even manage to create small lakes as a result.

However, due to the spreading of humans, unregulated courses of rivers that are required for this purpose have become rare. Therefore, beavers were almost extinct in Europe as a result of hunting and the loss of their habitat. However, they have been reintroduced in some areas for a few decades.

Humans shape their world

Ever since the so-called Neolithic revolution, when humans became settled and started to systematically engage in agriculture, the effects of humans' interventions on Earth can no longer be ignored. While before there were indirect effects — for instance by changing the population of huntable big game, which in return affected the development of vegetation — now, the direct influence of humans on

↑ Courses of rivers in silt
54°41' N · 8°46' E (July 15, 2002 and May 15, 2000)

From the Netherlands and Germany to Denmark, the coast of the North Sea is characterized by the constant turning of the tides. In the shallow bank area that is covered in water during high tide (right), the sea retracts for miles during low tide (left) and leaves the silt of the mud flats that are streaked by strongly shifting drainage channels, so-called tide ways.

their environment was apparent. Forests were cleared for cultivable land and pastures, but also for obtaining wood for wood charcoal and construction lumber. Larger and larger areas were used for growing selected cultivated plants. The construction of settlements and roads required additional space. By building dams and channels for irrigation and shipping traffic (in Eastern China, for example), the hydrological balance was influenced over a wide area.

While the effects of humans' actions on the environment only slowly increased over millennia and (apart from some of the aforementioned examples) hardly resulted in severe consequences for the environment, this started to change rapidly with the beginning of the Industrial Revolution a little over 200 years ago. The population growth resulted in an increased requirement for resources that affected many areas at the same time. This process has since continued at an accelerated tempo. Two hundred years ago, the world population was at approximately one billion people, but at the beginning of 2014, there were 7.2 billion people on Earth, each altering their environment at an increasingly rapid rate. Agricultural areas, the mining industry, growing cities, a multitude of all sorts of infrastructure facilities — almost everywhere on Earth, the shaping effect of humans is noticeable in the landscapes.

1,000 years are like one day

As is apparent from the different processes that shape the surface of the Earth, very different time scales can be found in this context. They range from a few seconds, in the case of blasting for road construction or a rockslide, to millions of years, in the case of the formation of mountains.

Apart from external influences, for example meteorites, and those from the interior of the Earth, such as tectonics and volcanic activity, all forces are a direct or indirect result of radiation from solar energy. This also notably affects the landscape-shaping effect of water, whose global cycle, as well as the annual freezing and thawing, are determined by the sun.

Multitude of forms

Besides the river landscapes, along with the valleys cut into mountains and alluvial plains, as well as meanders and mouths of rivers, the multitude of landscape forms that have been created by water include especially coastal landforms and the glacier landscapes of the high mountains and polar caps of the Earth.

In particular, the multitude of valley configurations looks amazing. Trough valleys in round shapes, ground by glaciers, with accompanying detritus layers from the moraines, and ravines that were sharply cut into rock by streams and rivers are the most common configurations to find in mountain regions. In flatter regions, those are meandered rivers through wide valleys filled with detritus in which a riverbed can easily change its course, especially during floods. Also, ground that was originally mainly flat can be cut through by miles-deep, bizarrely shaped canyons, as is the case with the Grand Canyon. With different levels of steepness of slopes, valleys are generally unstable. This results in more or less frequent landslides, which can change the valley landscape significantly by relocating and damming up the river. This can often create soil for the growth of a vegetation layer.

The landscapes of seacoasts, which can go through very different development processes, are similarly varied. The coasts of Norway and southern Chile are formed by sunken mountains, from the secondary summits of which only the tips rise from the water, while the flooded valleys reach deep into the country in the shape of fjords. In many regions, flat graded shorelines, formed by the sea currents running along the shore, with their characteristic lagoons and sandbanks can be found. An example of this type of coast would be the Baltic Sea coast of Poland with their lagoons. Wonderful examples for the variability of Earth, mirrored impressively in the images from the Earth's observation satellites, are the steep coasts in the South of England, or on the island of Ruegen in the form of chalk rock, that consist of relatively soft rock that is constantly being worn down by the sea.

Earth constantly recreates itself

Resulting from the constant interplay of forces, landscapes are continuously transformed, destroyed and recreated. This also involves permanent reshaping of habitats on Earth, including those that are used by humans. The processes that the water cycle is based on lead to the essential water constantly being distributed over Earth again and again, driven by the energy of the sun. In addition to the short-term changes that affect humans directly, the processes that run over millions of years and that are not noticeable are of importance. The creation of fertile soil and deposits of mineral oil and coal would be unimaginable without long-term erosion and sedimentation processes and without continental drift; the same applies to the creation of deposits of many other resources that are used by humans.

The river deltas of the polar zone are extraordinary landscapes. At the Canadian Mackenzie, only a thin layer of permafrost thaws in the summer and transforms vast areas into impassable marsh, crossed by individual lakes that are often given interesting geometrical shapes due to the alternations of freezing and thawing.

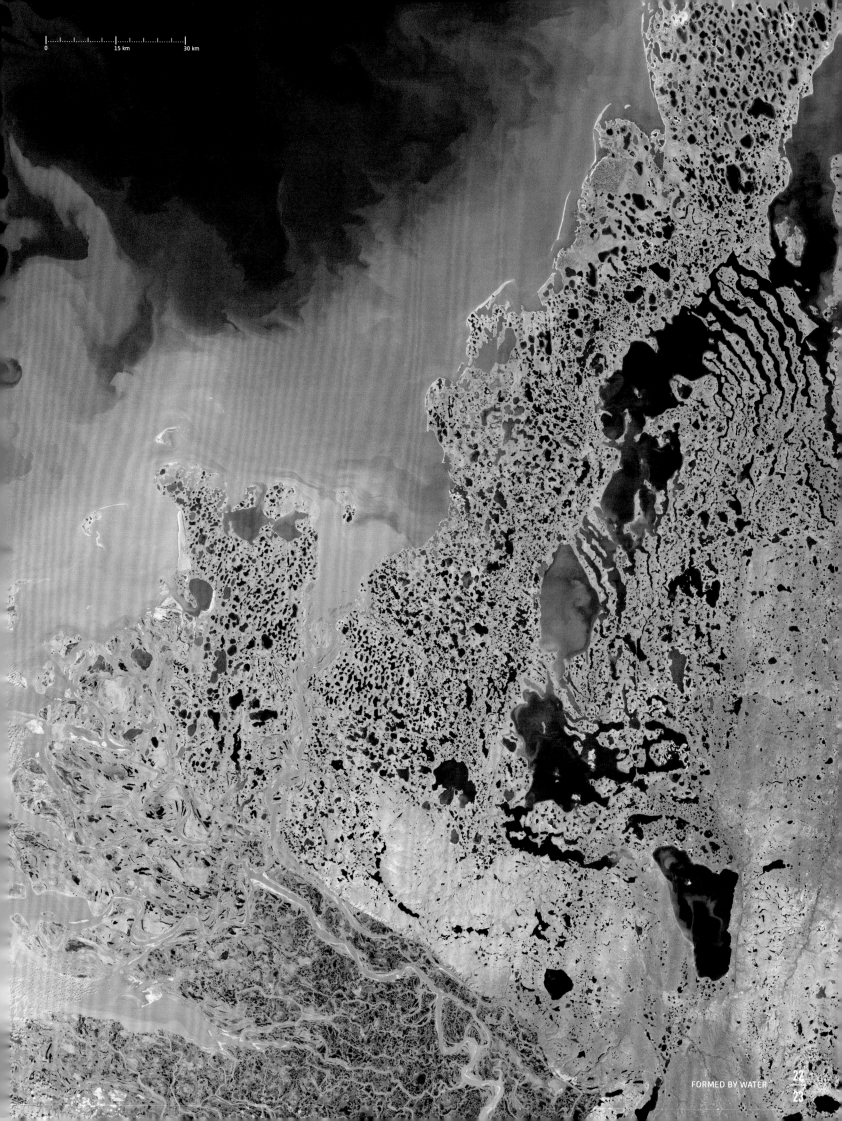

FORMED BY WATER

Traces of ebb and flow

39°05' S · 62°08' W (August 7, 2009)

At low tide, the light silt areas of the tidelands that have been formed by ebb and flow are exposed in the nature reserve Bahia Blanca at the Patagonian Atlantic coast and a finely branched out pattern of drainage channels appears. As part of a nature reserve, Isla Trinidad offers protection to dolphins, beaked whales and a colony of sea lions.

0   250 m   500 m

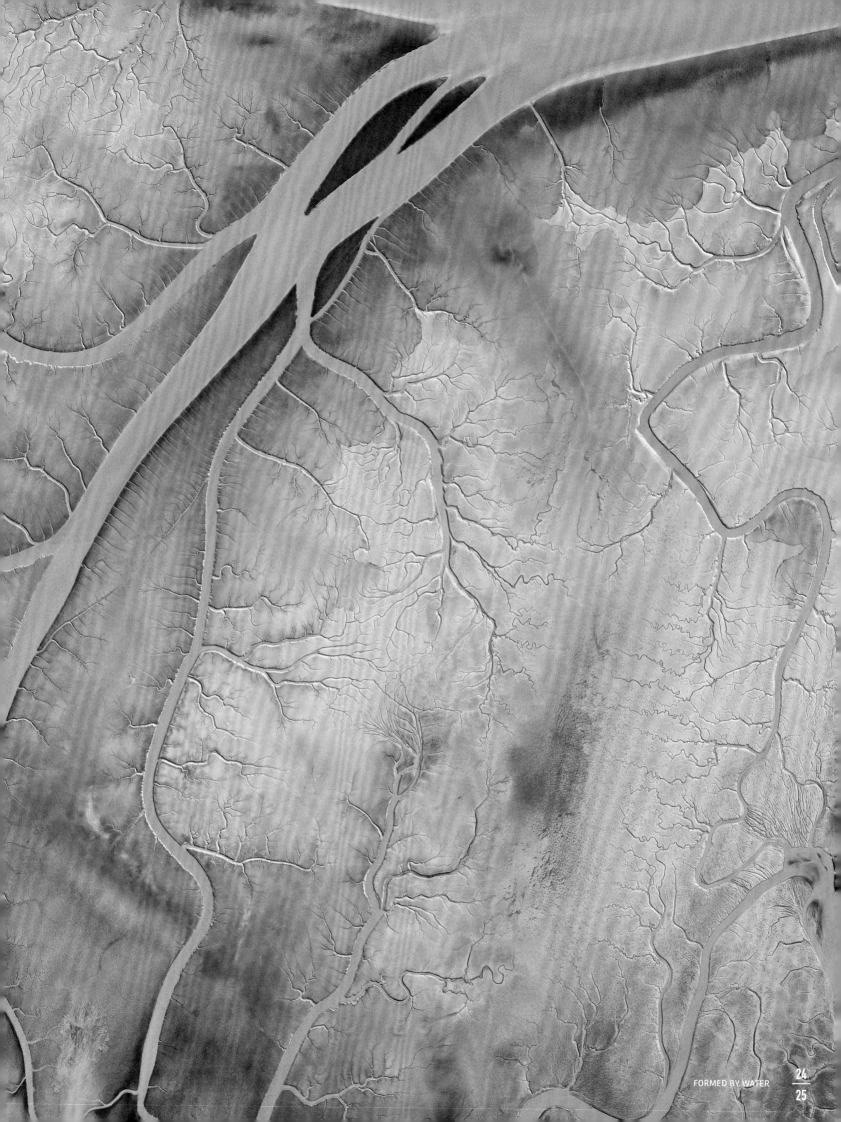

What is left by wind and water
25°46' N · 46°22' E (December 30, 2012)

Even if the water is not immediately visible, its effect is shown in the landscape, like here in the wadi al-Atk at the edge of the ad-Dahna desert north of Riad. For one, the remainders of the limestone plateau, the slopes and valleys of which were formed by run-off water, suggest that water has had an effect. On the other hand, the vegetation along the mostly dry courses of rivers of the wadis indicates the existence of water.

0   200 m   400 m

Folded
12°11' S · 134°07' E (October 16, 2009)

The Mann River flows into the Arafura Sea in the northwestern Australian Arnhem Land. Thick mangrove forests line the distinct meander bends in the flat estuarine valley. As a settlement area for Aborigines, Arnhem Land is usually barred to tourists; special permits must be requested up to a year in advance.

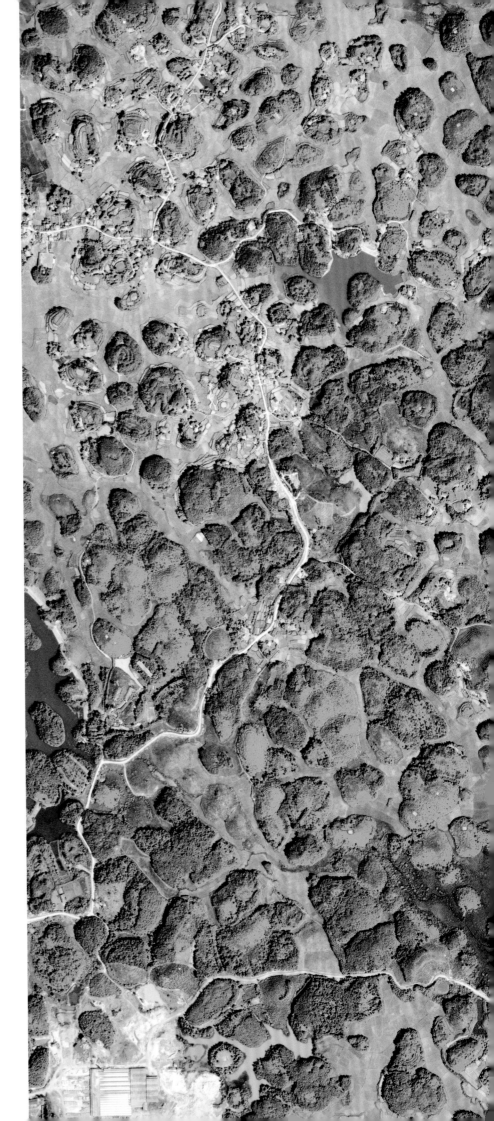

Spectacular rocks

21°08' N · 107°22' E (November 21, 2009)

The karst cones at the coast of the Halong Bay in Vietnam form a picturesque landscape. Approximately 2,000 limestone rocks that have developed as a result of weathering of rock due to water containing carbonic acid often soar over a thousand feet toward the sky or rise from the sea in the form of islands. The islands, with their grottoes and dripstone caves, are a UNESCO World Heritage site and a high-ranking tourist attraction.

HUMANS FORM THEIR WORLD

Ever since the Neolithic Revolution, with which agriculture and the associated settled lifestyle asserted itself, the traces of humans in landscapes became obvious. Even before that, hunting for large animal species (like the mammoth, which became extinct about 10,000 years ago) affected the landscapes since the extinction of the large herbivores resulted in changes in the vegetation. With the increase of the world population and the development of technical possibilities, the effects of human activities also increased. Cities, transportation networks and agricultural areas cover the planet. Today, there is hardly a landscape on Earth without traces of human activity, the changes not always being intended. One memorable example is the deforestation of wide areas of the Mediterranean region, which resulted in the karst formation of these areas. Today, intrusions with similar negative, although unintended, consequences can be found to a considerably greater extent in the Amazon Basin, for example.

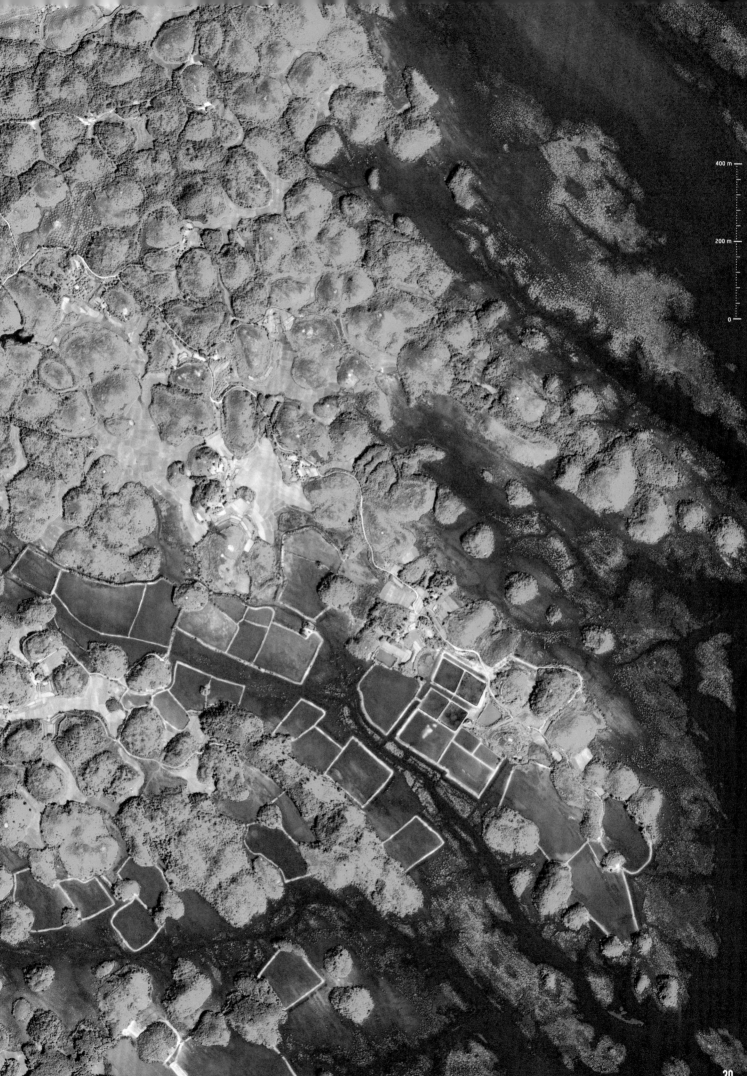

400 m

200 m

0

Between fire and water
22°10' N · 159°40' W (September 11, 2011)

The islands of Hawaii are characterized by the interplay of constructive volcanic forces and the erosion
forces of rain and the sea. This is also shown in the
Honopu Valley, which is difficult to access in the
northwestern part of the island Kaua'i. The dark volcanic rocks form a steep, towering coast that is shaped
by numerous valley cuts, at the slopes of which you
can find thick ferns.

From streams to rivers

30°08' N · 78°01' E (September 19, 2005)

The erosion patterns of both sides of the water-
shed in the Indian national park Rjaji are a clear
indication that the monsoon rains mainly come
from the south. Here, the network of drainage
channels is clearly more pronounced. The
remainder of the originally far-ranging sal forests
of the southern foothills of the Himalayas is
also the habitat of tigers, Asian elephants and
Indian leopards.

WHEN RAIN CREATES MOUNTAINS

As for the question regarding what influences erosion
and thus the creation of landscapes, climate plays one
of the main roles. Climate is responsible for water even
being available in any noteworthy quantity and whether it
appears in liquid form or as ice. Accordingly, the monsoon
climate, approximately eight million years ago, played
an important role in the development of the Himalayas
since the strong rain falls during the monsoon season triggered landslides and rockslides. The materials that were
transported when the deep valleys eroded were mainly
deposited in the Indus Delta and the Ganges Delta. Thus,
the weight of the mountain was also reduced, which in
turn led to its raising. The erosion that was triggered by the
monsoon rains was also an important contributor to the
impressive height of the Himalayas.

Worn down by wind and water

76°31' N · 25°01' E (July 31, 2009)

The long, drawn-out island of Hopen is located in the southeast of the Norwegian island group Spitsbergen (Svalbard). It is made up of table mounts that are rising from the sea, the slopes of which are deeply shaped by the erosion gullies of the run-off water. During winter numerous polar bears stop here on their journeys across the polar ice.

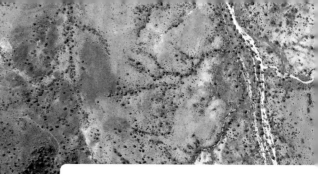

Knocked down to rubble

25°13' S · 134°34' E (September 26, 2010)

There are only small remainders of the original plateau left at the edge of the Australian Simpson desert
between the Eyre Lake and Alice Springs. Numerous
courses of rivers pointing in all directions and the
vast areas of detritus along the brim of the plateau
indicate the route the material took.

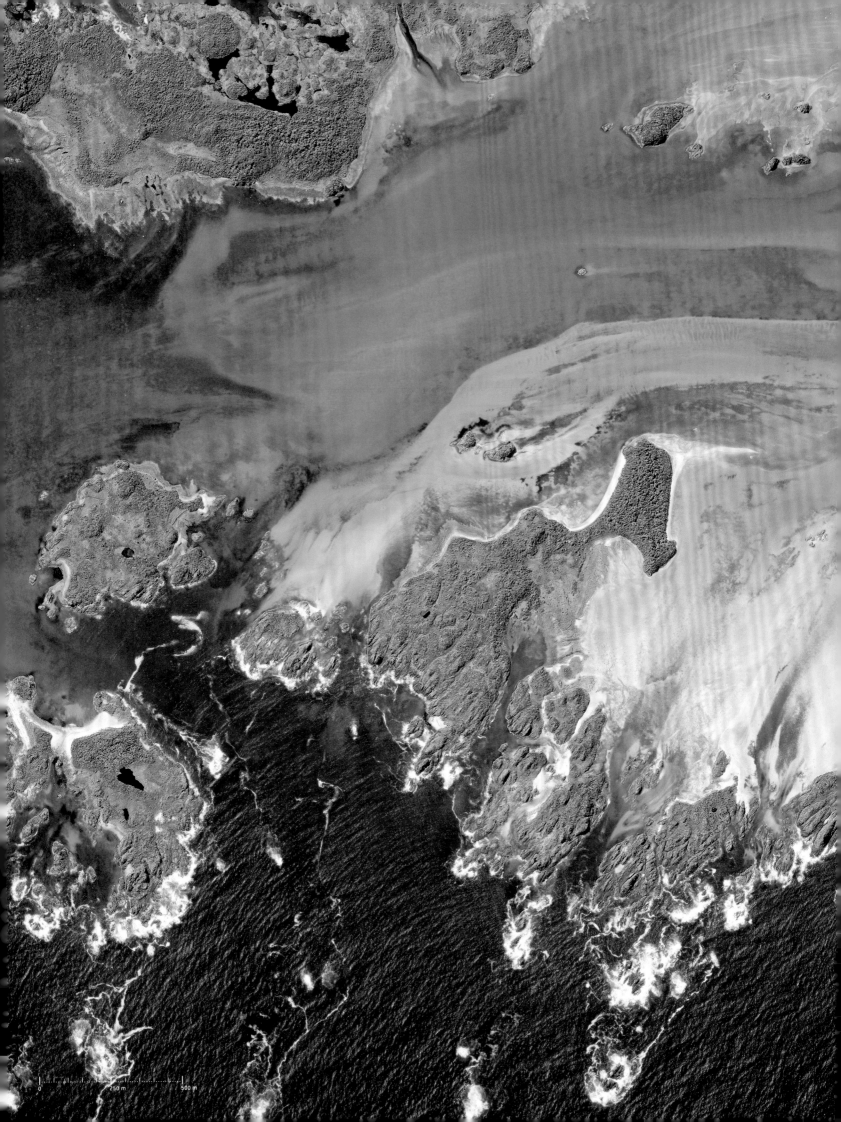

In the middle of the surf and tidal currents
48°55' S · 75°38' W (February 28, 2005)

Being very rugged and exposed to a rough climate,
the south of Chile is sparsely populated. The
Patagonian Archipelago along the Pacific coast
owes its structure to the subsidence of mountain
landscapes formed by glaciers. Esmeralda Island, in
the south of the Golfo de Penas, is one of these rises
that was created due to the tectonic activities along
the Pacific Ring of Fire.

Chinese arid zone

37°39' N · 76°44' E (February 26, 2002)

The foothills of the Kunlun Shan still reach 9,850 feet (3,000 m) in the transition area to the Tarim Basin. The basin is one of the areas in the world with the lowest precipitation. The incidence of light and the snow that remained on the shady slopes let the finely ramified drainage pattern in the soft, almost bare layers of sedimentary rock clearly stand out.

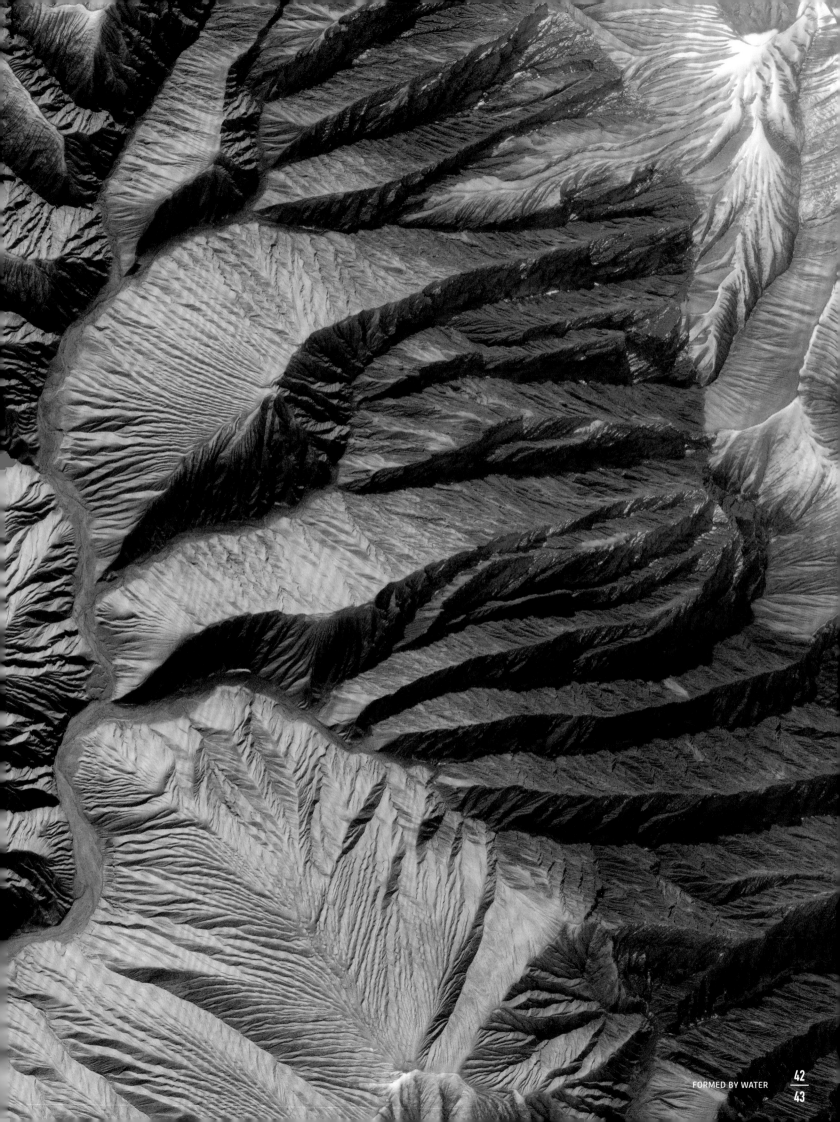

0   6 km   12 km

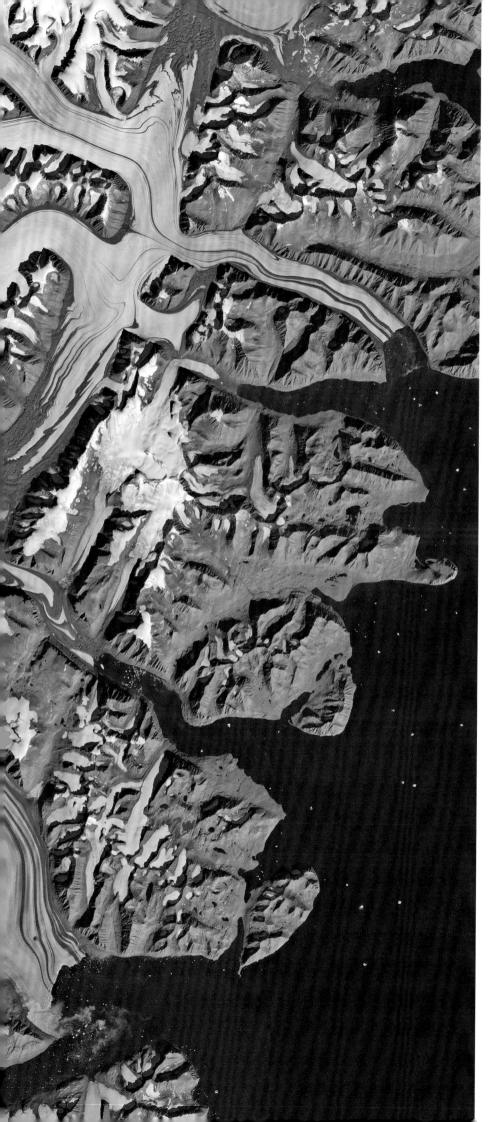

Frozen rivers
69°37' N · 25°18' W (August 28, 2000)

In Greenland, the ice streams of the glaciers stretch across valleys that they formed themselves over many miles, the complex stream lines standing out due to the dark medial moraines. The weight of the ice crust presses the subsoil down so that entire mountains are below sea level. Once the weight lifts, they rise again.

GLACIAL HERITAGE

When you look at one of the large alpine glaciers that lie in their valleys like frozen ice streams, it is hard to believe that glaciers are extremely efficient shapers of landscapes. Although they flow only very slowly, they have a severely eroding effect on the subsoil due to their weight and detritus that they carry along. These effects become visible when the glaciers retract, as was the case at the end of the last ice age. Even today, many of the ground valleys form the basin of lakes, for example in the foothills of the Alps.

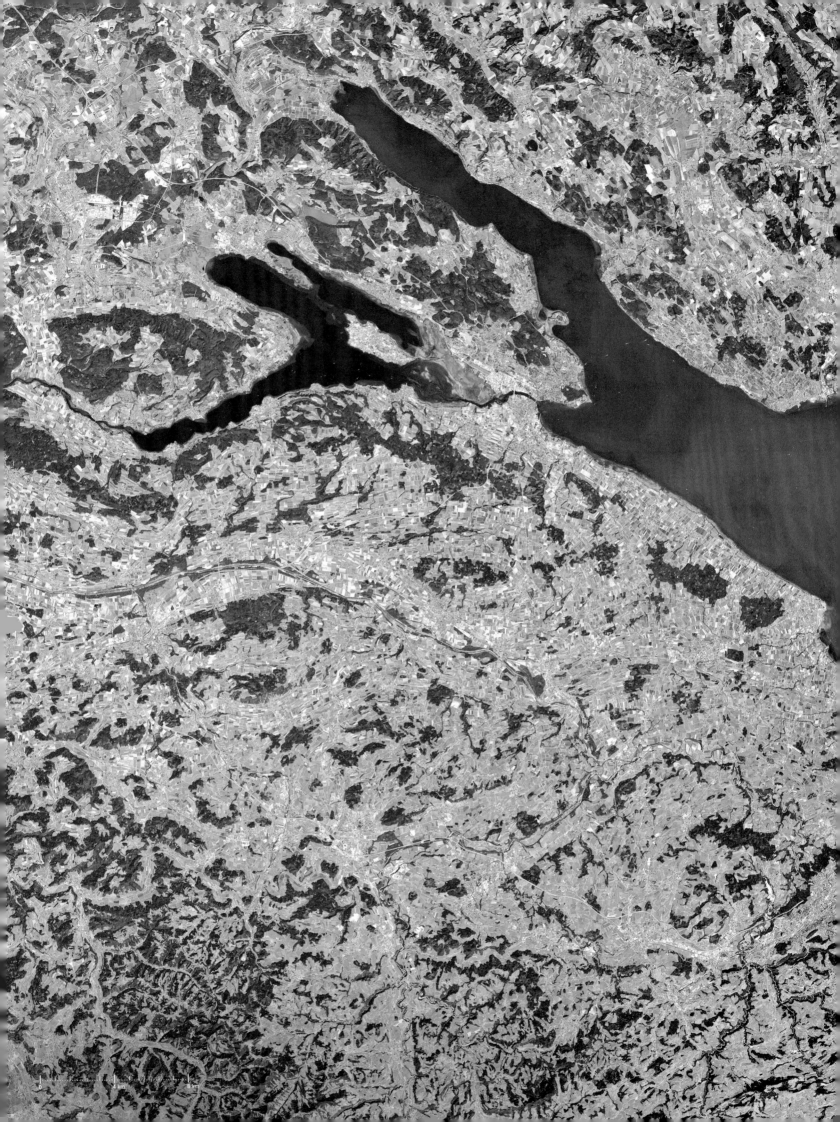

Heritage from the Ice Age
47°35' N · 9°24' E (August 1, 2001)

Lake Constance is a typical representative of lakes in the foothills of the Alps that was formed by the last ice ages. The previous Rhine Glacier ground the valley, which filled with melt water approximately 16,000 years ago. Not even Lake Constance is permanent. Bit by bit, it is filled by sediments from the Rhine, until after a phase of marsh, it will completely turn into land and create a river basin.

Deep cuts

36°01' N · 113°09' W (September 2010)

In the southwest of the United States, the Colorado River has created a unique landscape — the Grand Canyon. During the course of some ten million years, the river dug canyons up to 5,900 feet (1,800 m) deep into the previously flat sedimentary plain. Since they carry significantly less water, the tributary rivers of the Colorado were not able to cut that deep into the rock.

0   7.5 km   15 km

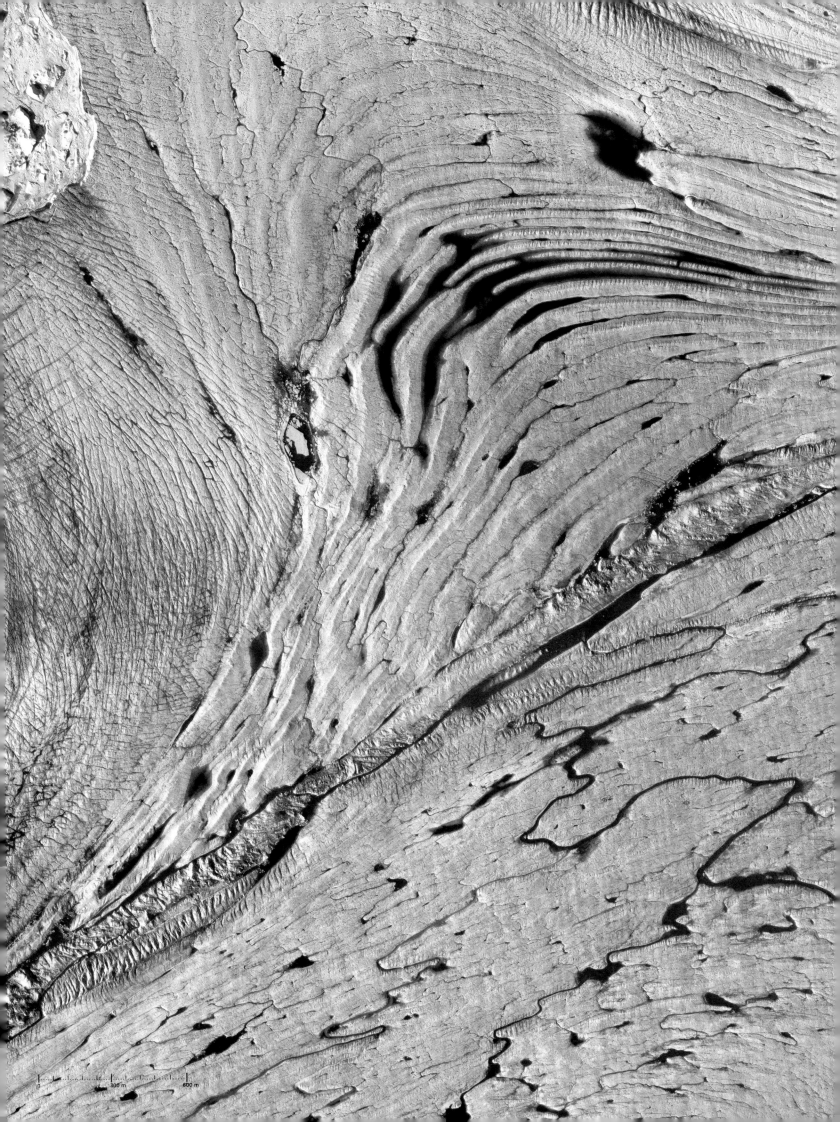

The forces of ice and water

79°34' N · 21°51' W (July 9, 2012)

The Nioghalvfjerdsfjorden glacier in the northeast of Greenland is located at at latitude of almost 80° N and is only about 685 miles (1,100 km) away from the North Pole. The fjord that was previously formed by a glacier is now filled with water in which the ice of the glacier floats. In the summer, the melted water on the glacier surface forms small lakes and streams that cut into the ice and glow in shades of blue.

Rivers of sand
24°34' N · 32°21' E (August 9, 2010)

At Edfu, the Egyptian Nile Valley is demarcated by
a cliff from the desert in the west. Old wadis, filled
with desert sand that originally drained the hilly land
up to the Nile and now have the character of sand
rivers, stretch through the cliffs that have been worn
down by the wind. Erosion has clearly carved out the
geological fault lines in the rocks.

0   250 m   500 m

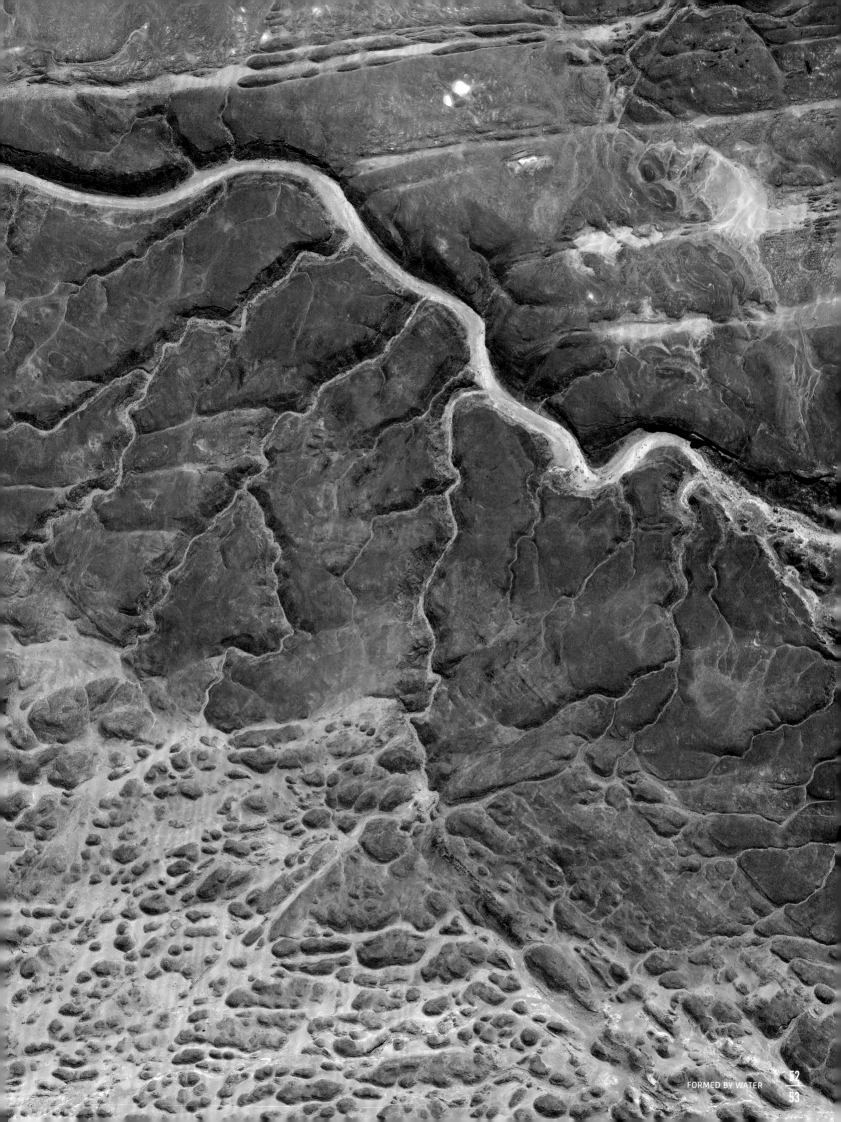

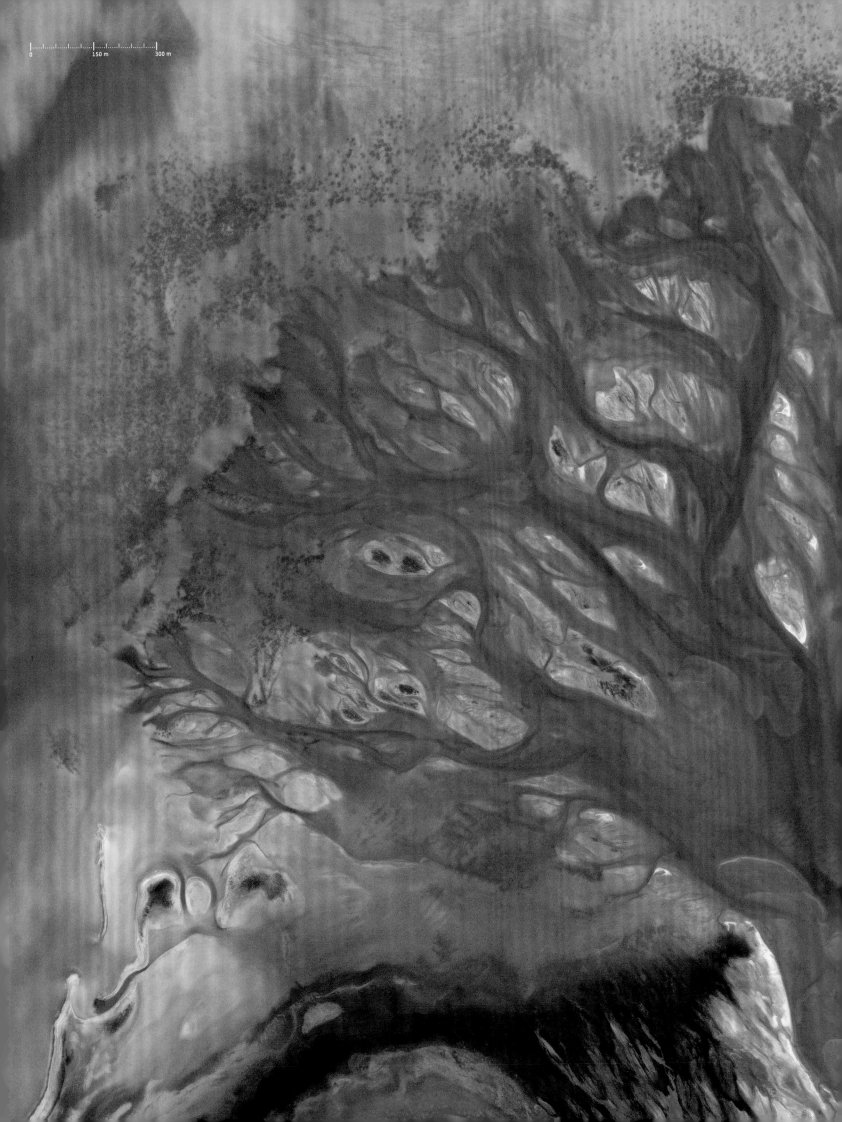

A highly ramified alluvial fan marks the transition between two partial basins of the Khor al-Adaid, also called Inland Sea. The lagoon-type inland lake in the desert southeast of Qatar is connected to the Persian Gulf via a natural channel. It is therefore also exposed to the surf and tidal currents, which shape the debris cone. At the same time, Khor al-Adaid is important as old cultural land and as a nature reserve.

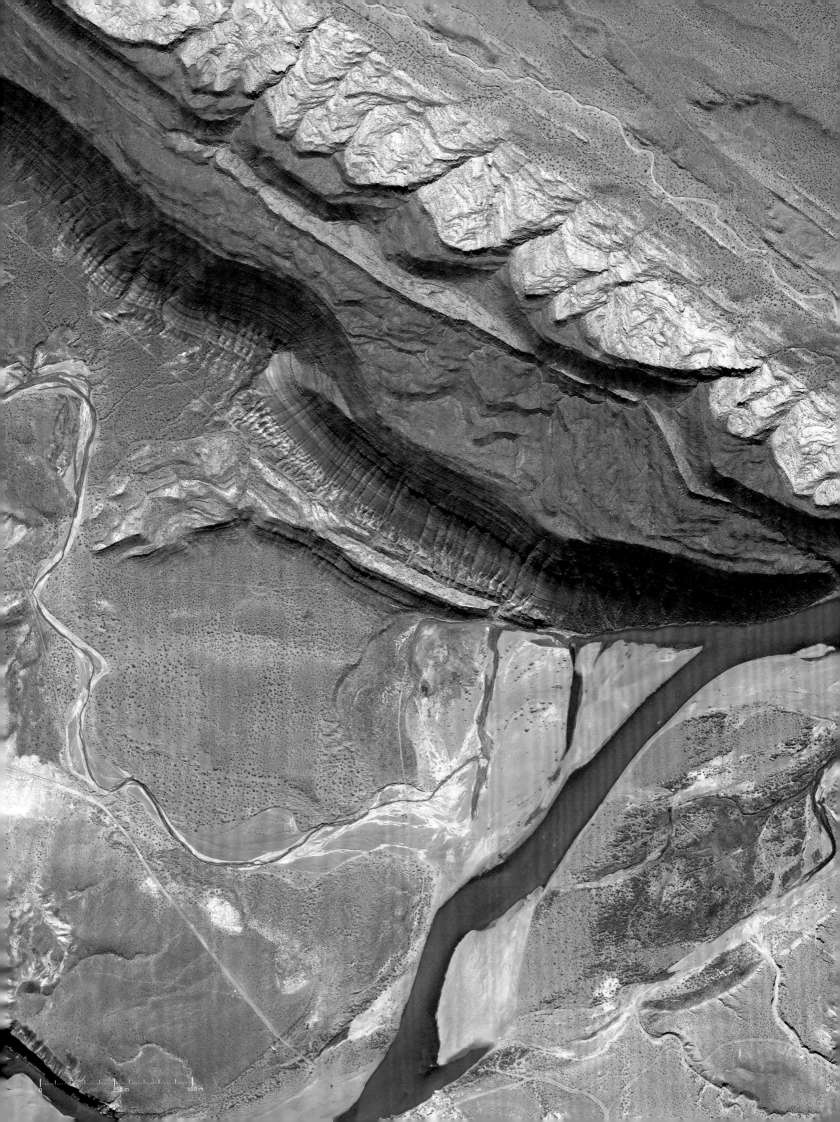

Deep cuts

37°40' S · 69°57' W (September 17, 2005)

On its way from the Andes Mountains to its mouth
into the Río Negro, the Río Neuquén negotiates more
than 6,560 feet (2,000 m) in height. Over many years,
the river has cut deeply into mountain barriers that
had it dammed up to a large lake, originally. Vast
areas of rubble attest to the incredible strength of the
river during the snowmelt.

Ice landscape
62°46' N · 151°17' W (August 9, 2003)

The Kahiltna Glacier stretches from the highest mountain in North America at 20,321 feet (6,194 m) to sea level at 1,000 feet (300 m). A number of side glaciers flow into the 37-mile (60 km) long ice stream, the ice bodies of which can be identified over many miles by the rubble of the medial moraines. In late summer, the dark bare ice is exposed and melt water creates small lakes.

0   350 m   700 m

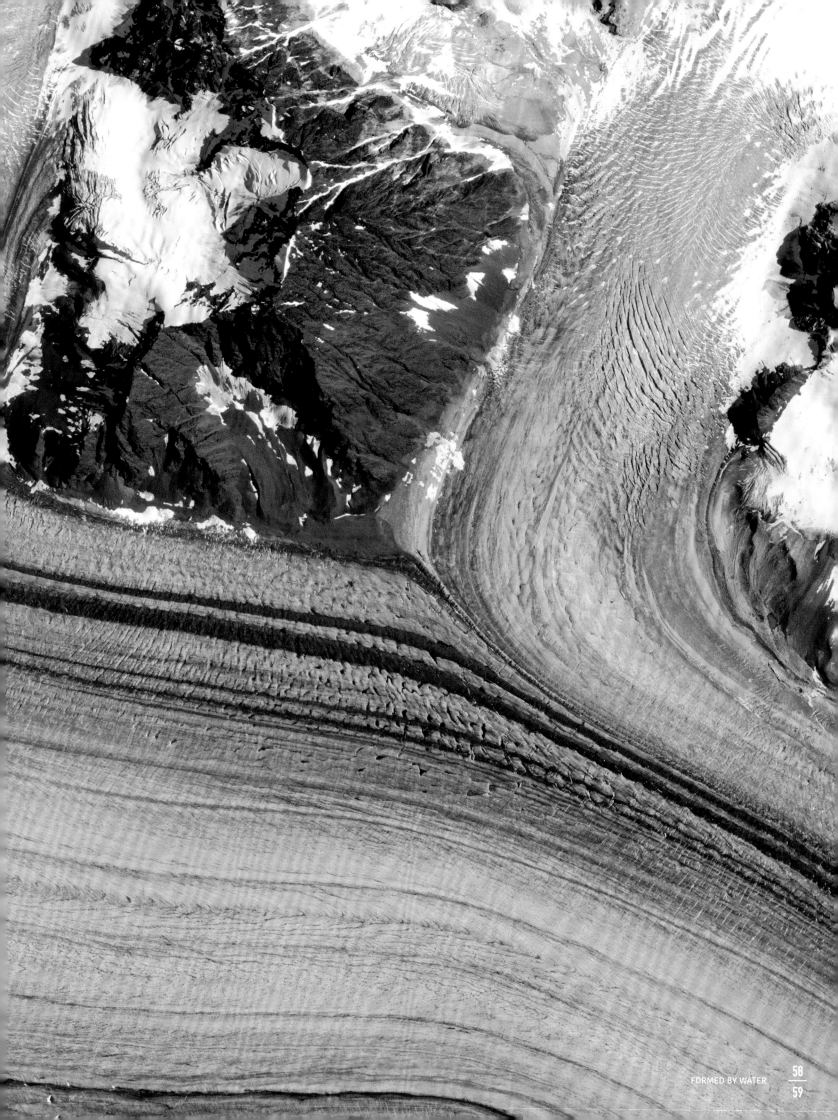

GEOHAZARD "HUMAN"

How climate change altered the face of the Earth

Over millions of years, the climate on Earth has significantly changed time and time again. Ice periods and warmer periods alternated. However, the warming process has never been as rapid as it is today.

Three large creative powers form the face of our planet: plate tectonics, climate and life. The drive for plate tectonics, in other words the continental drift and creation of mountains, lies in the hot interior of Earth. The energy stream that drives climate and life, however, comes from the sun. Life on Earth is no passive passenger that has just adjusted to the conditions on Earth. Contrary, life has already taken over our planet for billions of years and has completely redesigned it. Climate conditions are also shaped by life. Every hiker is aware of that on a small scale: The climate in the protection of a forest is completely different to the climate on the adjoining barren detritus area. On a large scale, the vegetation cover significantly changes the brightness of the Earth's surface and thus the level of solar irradiation that is reflected into space — and, in return, the thermal balance of our planet. And what is even more important is that life has completely changed the gas composition of our atmosphere. Without life, the air would basically be free from oxygen. But also the quantities of the important greenhouse gases carbon dioxide and methane are pivotally controlled by life, and thus the climate. For example, tiny microalgae in the oceans, the lime shells of which sink to the bottom of the sea after their death, play a central role: This deposit in the sediments is undoubtedly the most important process that contributes to the removal of carbon from the climate system. It balances the quantities that are added due to volcano action. As a result, scientists consider the biosphere an integral part of the climate system — we would therefore not even consider climate and life as two separate creative powers. Climate and life develop and change jointly — and shape the face of our Earth in the process.

Insights into the large climate machinery

Deposits at the bottom of the sea, or greenhouse gases in the atmosphere, are important aspects of the large climate machinery that are not visible to us. The most immediate visible signs of large changes in climate, on the other hand, are the ice masses of Earth. In the modern era, satellites send us spectacular images of glaciers, pack ice on the sea and gigantic ice sheets and their changes. We know that our planet has been gripped by cyclically recurring ice ages over the last three million years — warm periods like the past approximately 10,000 years, were an exception during this phase. You must imagine the ice masses at the peak of these ice times as gigantic: Ice layers that were several thousand feet thick and that were created by snowfalls covered large areas of North America and Eurasia. At the time, there was approximately three times as much continental ice as today. And the coastlines also looked noticeably different because the global sea level was approximately 400 feet (120 m) lower than today, since the water missing in the sea was lying on land in the form of ice. If you go back further in the history of climate, to approximately the past 500 million years, you will find almost no ice on land most of the time. Mostly, the climate was considerably warmer, in all probability because of larger concentrations of carbon dioxide in the air, which is connected to very slow, natural changes in the aforementioned carbon cycle. So, the Earth's history shows: The climate system is by no means unlimitedly stable.

About the disappearance of the arctic ice cover

From a climate history point of view, we are currently in a rapid global warming phase. Since the late 19th century, the global temperature has risen by one degree. As a comparison, at the end of the last ice age, (approximately 15,000 years ago), the global temperature increased by around five degrees, but this increase took approximately 5,000 years — thus, it was at a rate of only about 0.1 degrees per century. The main cause for this modern global warming is the human species. We know this for example because we understand the thermal balance of our planet well enough. The largest disturbance for this thermal balance is the increase of the greenhouse gases in the air, mainly the carbon dioxide. Its main source is the burning of coal, mineral oil and natural gas — as a result, today, we deliver about 100 times as much carbon from the Earth's crust into the atmosphere per year as originates on average from all the volcanoes. The carbon dioxide percentage in the air today is one-third higher than it was for at least a million years. If we do not stop global warming very quickly now, within the next decades, it will catapult us far out of the experience range of the stable climate period of the past 10,000 years, during which the human civilization was able to develop. The face of our Earth will change considerably. Different to the changes of previous Earth history, this time the cause is humans — and it will happen dramatically faster. Already now, the occurring changes are clearly visible from space. The summer ice cover on the Arctic Ocean has shrunk to half over the past three to four decades. And ice on land melts, too — in the Alps, about half of the glacier ice has disappeared during the past hundred years, and recently this shrinkage even accelerated. Large ice sheets in Greenland and Antarctica are also shrinking. Measurements from satellites document: The sea level has risen by over an inch (3 cm) per decade during the past decades. This rise of the sea level that will accelerate even more if global warming continues will continuously change our coastlines in the course of the next centuries. So far, the sea level has risen worldwide by approximately 8 inches (20 cm) since the 19th century; before then, it had been stable for thousands of years. Already today, the effects are noticeable at some locations in the form of disastrous storm surges. In this century, the Intergovernmental Panel on Climate Change anticipates a rise of the sea level of up to 3 feet (1 m) — in the next centuries, several further feet will more than likely follow if the warming is not stopped by resolute political action soon. Entire island states will sink and some coastal cities will have to be abandoned.

Will we be able to live with the consequences of global warming?

However, the face of our Earth will also change by the effects of climate change on the cover of vegetation of our continents. Some regions (e.g., the Mediterranean area) will be challenged with significant problems due to increasing drought — to some extent, they are already experiencing that. Forest fires are on the increase; large areas of forest are at risk of being lost. For us humans, the expected negative effects of climate change on crop yields that may significantly endanger the food security in some regions of the world are particularly serious. During its history, our Earth is a planet that has already undergone many changes, some of which were dramatic. And the modern climate change that is caused by humans will again significantly redesign the face of the Earth — we are only just at the beginning of this already clearly visible and measurable process. Earth will survive this change, too. However, the question is how well will humans cope with it?

Professor Stefan Rahmstorf, born in 1960, is an oceanographer and climate scientist. He teaches the subject "Physics of the Oceans" at the University of Potsdam, Germany. Rahmstorf served the Federal Government of Germany in the German Advisory Council on Global Change (WBGU) from 2004 to 2013 and is member of the Academia Europaea. He was furthermore one of the lead authors of the "Assessment Report from the Intergovernmental Panel on Climate Change" (IPCC) that was published in 2007 and has published over 90 professional publications and several books on climate change.

"The effects of global warming: Entire island states will sink and some coastal cities will have to be abandoned."

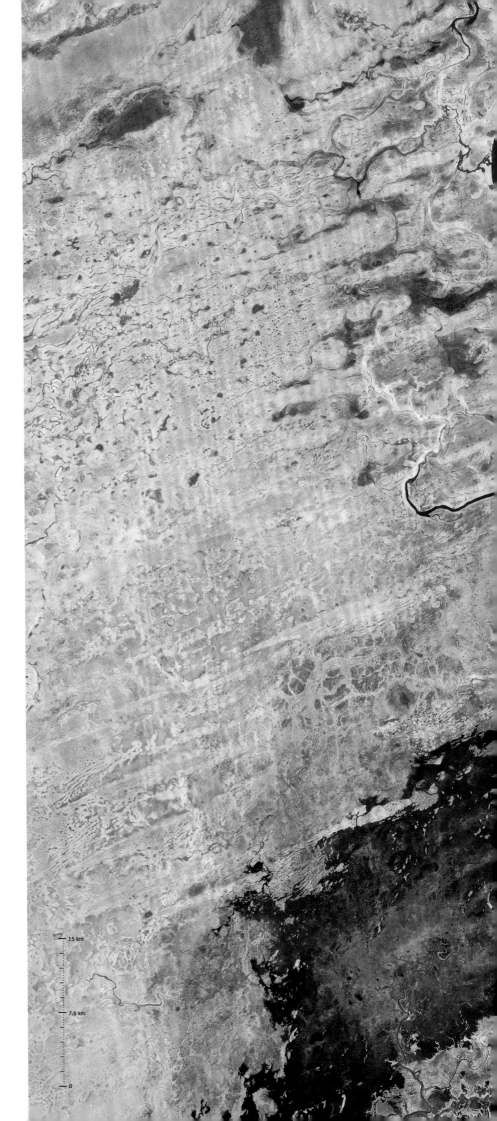

Floods and sand dunes

15°14' N · 4°14' W (October 19, 2002)

The inland delta of the Niger is in the Sahelian
zone in Mali. As extensions of the Sahara, long
bands of sand dunes stretch into the area. The
water level is subject to severe seasonal fluctua-
tions. From the areas that are located at the upper
reaches of the Niger, the water from the rainy
season needs several months until it reaches the
inland delta so that it is still flooded during the
dry season.

SHORT-LIVED BEAUTY

The landscapes of Earth are exposed to continuous
changes — also lake and river landscapes, where the influence of water leads to often particularly rapid changes.
Lakes, especially, experience constant transitional events.
In the course of time, their basins are filled by the sediment load of their feeders so that there is no space left for
water. When a lake comes to its end, far-reaching marsh
landscapes are formed and become increasingly more
dry over the years, until grassland or forest spread across
the area of the previous lake. The life duration of a lake
is dependent on many factors. This includes the climate
and the size of the lake and its catchment area, but also
geological conditions of the surroundings.

Crow's foot in the Gulf of Mexico
29°12' N · 89°41' W (November 2006)

The Mississippi Delta was formed from the sediments
of the river that were deposited here due to the low
currents in the Gulf of Mexico. As a result, the mouth
ends have moved further and further into the open
sea and have given the delta the shape of a bird's foot.
Regulating measures at the river have resulted in a
decrease of the sediment supply, so that the delta is
assaulted by the sea at the edge.

0   7.5 km   15 km

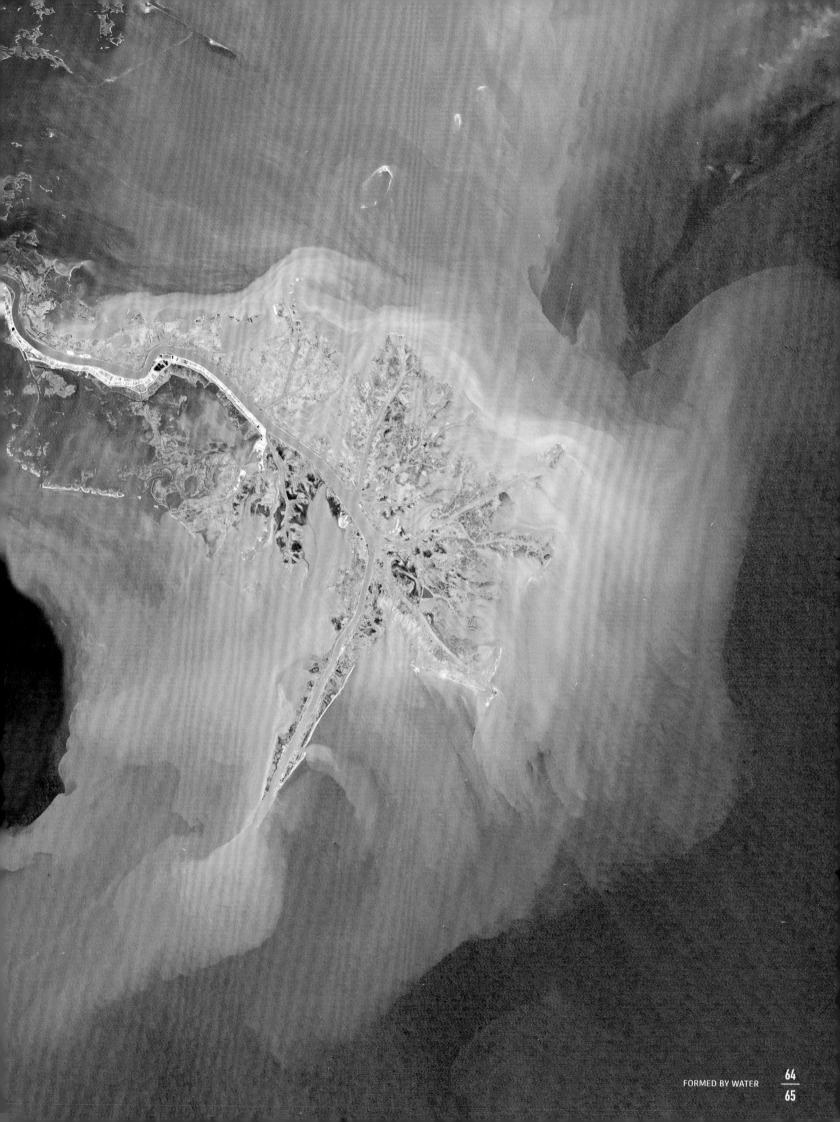

New land at the polar sea
72°59' N · 127°29' E (July 27, 2000)

With an area of 17,375 square miles (45,000 sq. km), the Siberian Lena Delta is one of the largest river deltas on Earth. Materials that are transported off by the Lena and its feeders are deposited here and constantly force the river to search for new routes to the sea. This results in the formation of numerous courses of rivers that push the edge of the delta in a fan shape into the polar sea.

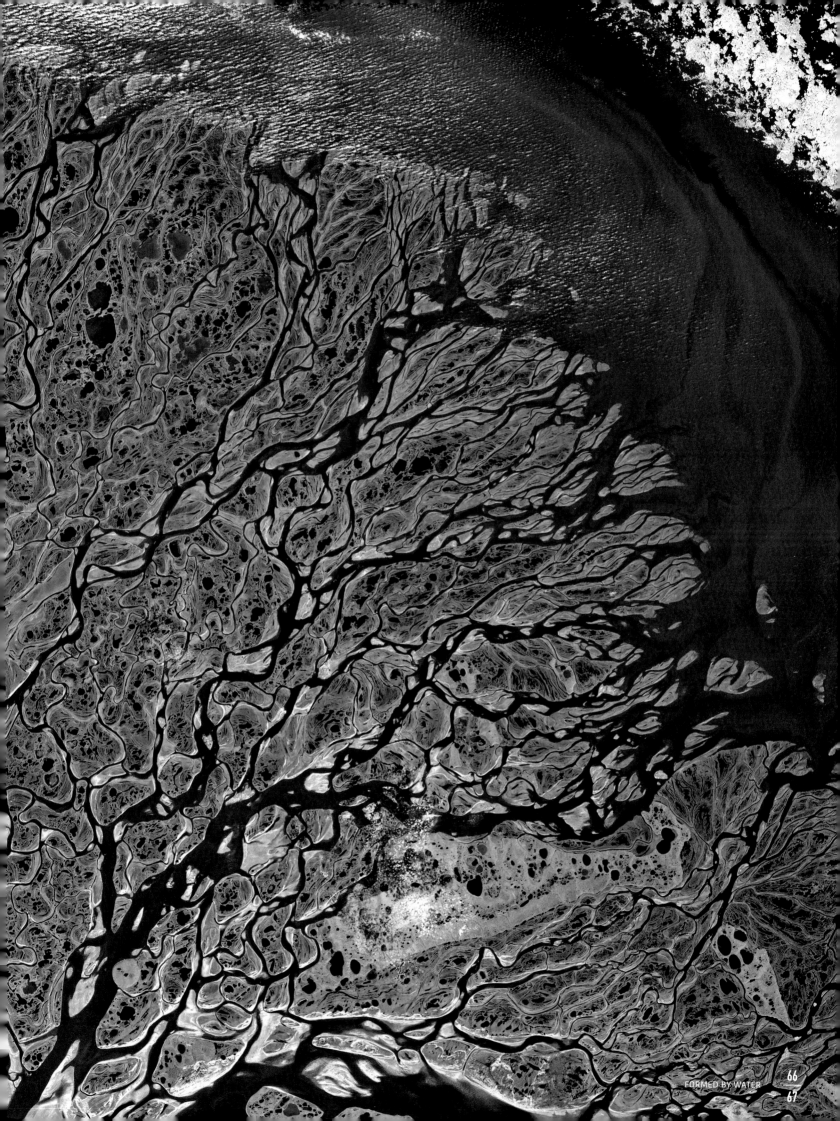

Enormous wall of water
17°56' S · 25°51' E (September 16, 2011)

At the border between Zambia and Zimbabwe, the Zambezi River sweeps over the Victoria Falls toward the valley. This waterfall, which is the largest in the world based on the product of the height of almost 360 feet (110 m) and the width of over 5,577 feet (1,700 m), is called quite appropriately "The Smoke that Thunders" in a local language.

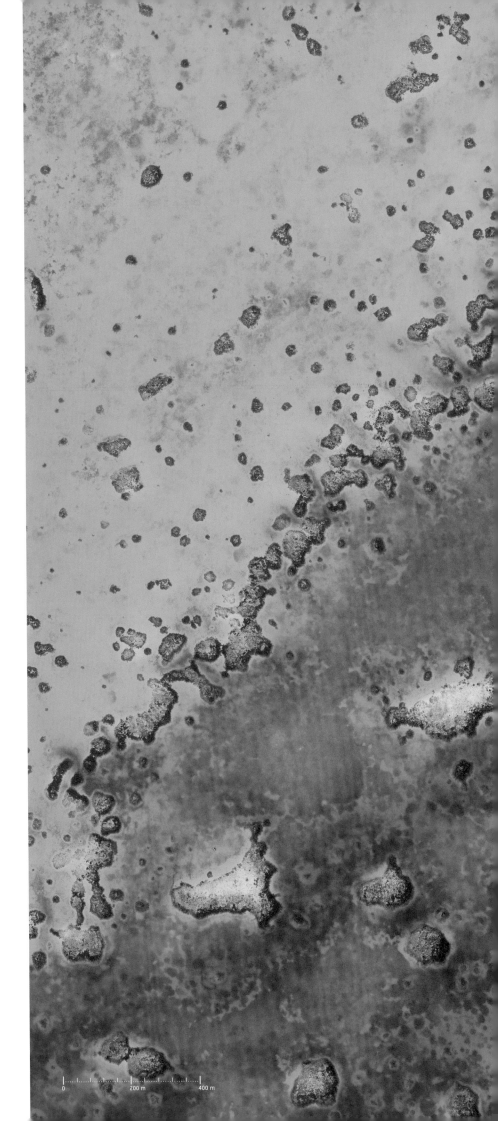

The Caribbean coast at Belize is surrounded by coral reefs and atolls. In the middle of one of these atolls, the Lighthouse Reef, you will find the Great Blue Hole, which clearly contrasts with its surroundings due to its dark color. The hole, that is almost circular with a diameter of approximately 985 feet (300 m), is the result of erosion forces — a cave was washed out from the limestone of the subsoil and was later flooded by the sea.

COMPLEX LANDSCAPE FORMATION

The formation and vanishing of landscapes is the result of processes that are often linked in a complex manner. The phenomenon "karst" shows this very clearly. Even the limestone that forms the karst landscapes is the result of a long history. It is usually formed by pieces from corals, shells and snails that have sunk to the bottom of prehistoric seas where they then solidified. Geological processes later raised these sediment layers and they formed the basis for massifs. The low weathering resistance of lime is the prerequisite for this rock forming special, sometimes bizarrely shaped structures. Limestone caves, dolines and the underwater version of the "blue holes" are examples of the varied landscape forms of the karst.

0   200 m   400 m

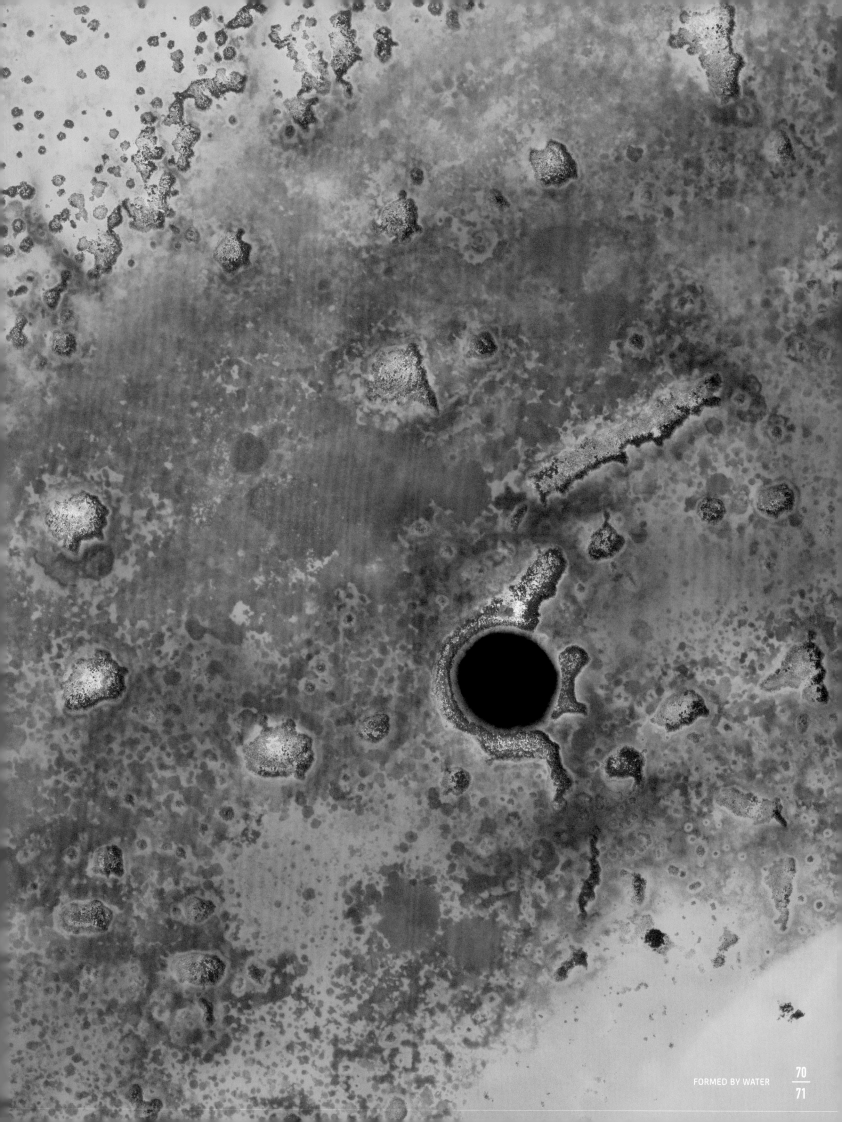

Where freshwater and saltwater meet
30°30' N · 48°53' E (August 14, 2013)

The wildlife preserve Shadegan near the Iranean city
of Abadan is only approximately 30 miles (50 km)
away from the Persian Gulf. In the area of influence
of ebb and flow, the river banks are clearly formed by
the runoff and backflow of water. With a size of almost
1,160 square miles (3,000 sq. km), the wetlands ex-
hibit a very special vegetation, which also makes them
a paradise for numerous species of bird.

Annual growth rings of the seashore
22°03' N · 105°38' W (November 13, 2009)

Like annual growth rings, the old courses of the coasts of the Pacific Ocean run closely side by side. In the Mexican State Nayarit, currents have deposited sediments at the flat coast and with it, extended the coastal plain bit by bit. The mangrove swamps that formed this way are interspersed with a thick system of lagoons that are also used for shrimp cultivation and are integrated into the nature protection area "Marismas Nacionales."

0   250 m   500 m

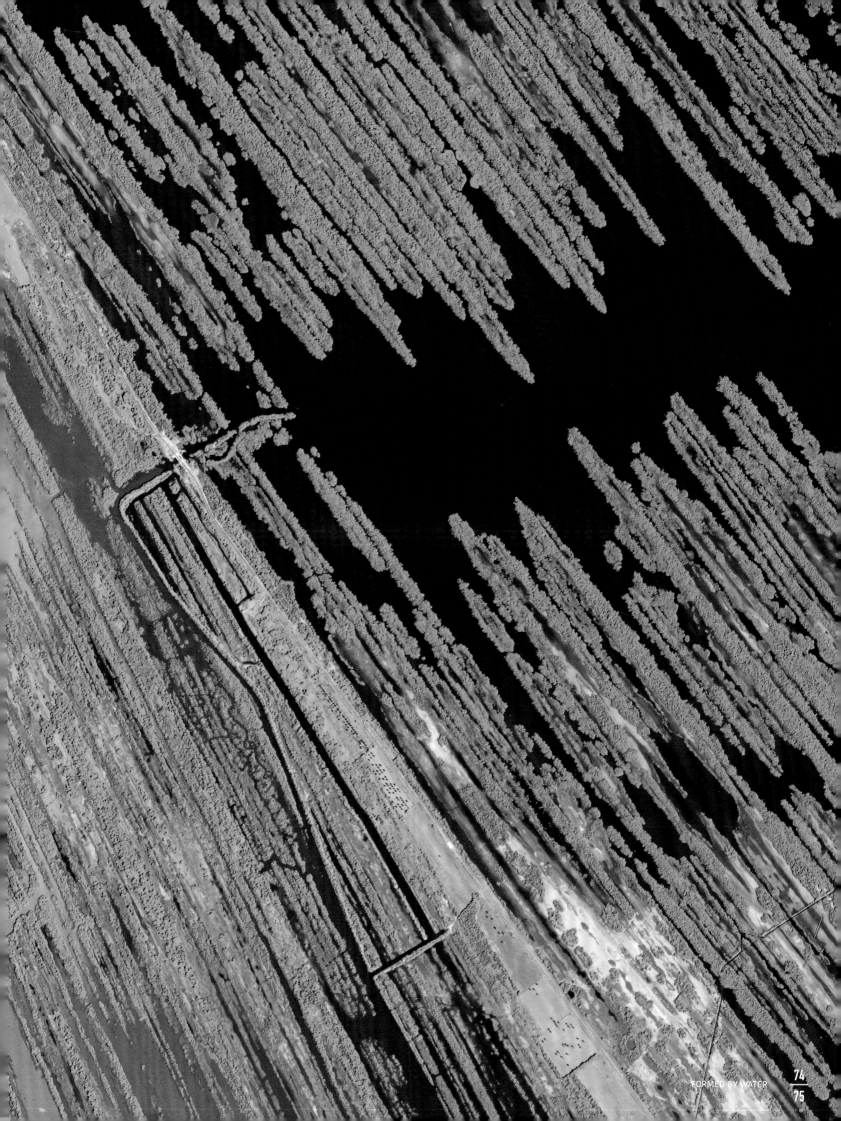

Colors of salt
33°32' S · 122°39' E (March 4, 2006)

In addition to the dry climate, agriculture in the southwest of Australia is also significantly challenged by the salinization of the soil. Vast areas of arable land are interspersed with salt pools that are colored differently by algae and that form light salt crusts when the water evaporates.

0   200 m   400 m

Island ornaments in the Mississippi
43°12' N · 91°09' W (October 12, 2010)

Protected by a dam that has been constructed upstream, solid wooded sandbanks have established at the upper course of the Mississippi between La Crosse and Dubuque, which form an intertwined ornament in the wide river valley.

Settlement area
"Water"
BUILT ON
WATER

Where there is water, humans can settle
A GOOD LOCATION TO LIVE

The Orkney Islands, with the stone-age excavation site Skara Brae, are located in the far north, roughly at the same geographical latitude as Stockholm, St. Petersburg and the south of Alaska. When you look at the stone structures, it is not immediately clear what strikes you more — the age of the buildings, which is estimated at approximately 5,000 years, or their location in the rough climate between the North Sea and the North Atlantic, where this settlement was created in close vicinity to the seashore. Besides signs of livestock breeding, excavations revealed numerous oyster shells and remains of fish, clear indicators as to the reason why this settlement had been built at this particular location.

In our day and age, satellite images have been available for some years that show the entire surface of Earth at night, at increasingly improved quality. These images highlight those areas where humans spread out with their activities, because mainly artificial light sources are visible, such as the illumination of roads and buildings. When you view these night shots, it is always astounding how easy it is to orientate yourself based on these images.

One important reason for this easy identification of the continents, for example, is the fact that numerous light points mark the course of the coastlines. Some courses of river also show similarly in the form of illuminated bands. Excellent examples are the valleys of the rivers Nile and Indus. The satellite images of Earth at night thus provide hints as to the important role that water plays for the settlements on Earth.

At quiet banks and shores

In some of the theories with which anthropologists try to explain the development of humankind, water played an important role. Based on one of these theories, water was a contributing factor to the development of the erect posture of man. In accordance with this theory, the ancestors of humans living by water have been fishing in the flat bank and shore areas and collected shellfish and mussels. An erect posture allowed

"The water is a friendly element to a man who is at home in it, and who knows how to deal with it." Johann Wolfgang von Goethe

← Lights in the dark

Center of the image: 48°00' N · 14°45' N

In addition to isolated oil rigs in the North Sea and Mediterranean Sea off the coast of Libya, mainly the settlement areas of humans emerge as illuminated areas at night. The areas of high population density that are often located at the coasts or in river valleys appear particularly bright. This presentation of the light sources of Europe was compiled from numerous individual night-time satellite images and combined with a background that was created from a regular satellite image map.

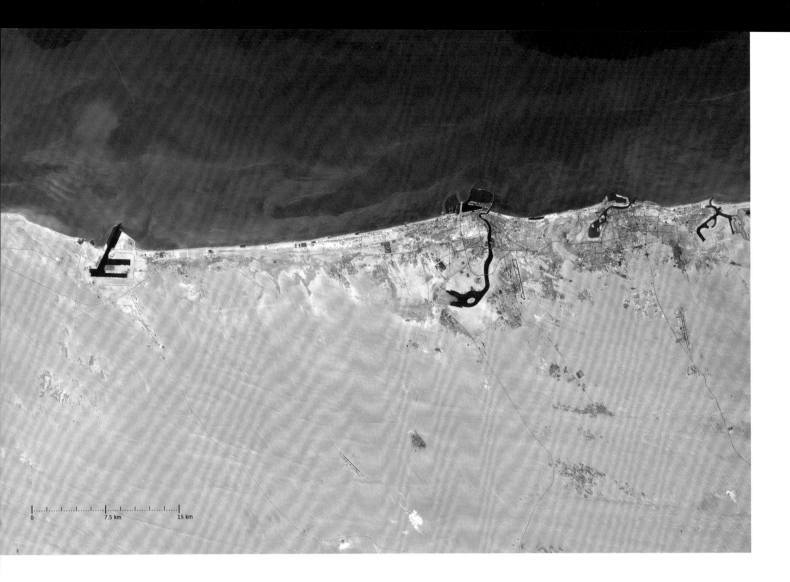

them to use their hands more efficiently and at the same time, to walk into deeper water where they could find more ample food sources.

Historians have found clear indications in all parts of the world, even for later time periods, that humans always had reasons to settle at or near water. This already applies to the oldest known settlements. The settlement ÇatalHöyük, in today's Turkey, which dates to the time period as of approximately 7400 BCE, for example, is located in the relatively dry Konya plain on the debris cone of a river that ensured the drinking water supply, and also formed the basis for agriculture operated near the settlement.

In later years, as the settlements grew to cities and the possibilities of using water became more extensive, the topic of "water" became more important. Soon, the irrigation of agricultural areas was added to the initial supply of the population with drinking water. One example that is particularly demonstrative is the valley between the two rivers Euphrates and Tigris, where fields were irrigated by means of a sophisticated channel system with numerous sluices in the third millennium BC. For many historians, the organizational requirement in connection with the construction and operation of these systems represented the starting point of creating the complex social structures of higher-developed countries.

Why at the water?

Surely the easy accessibility of drinking water and huntable animals, as well as the favorable conditions for blooming and agriculturally usable vegetation, are some of the original reasons to settle at the banks of bodies of water. Even today, this is evident in dry areas. One example that is clearly visible from space is the Nile Valley, which forms a small green band of life in the middle of the hostile Sahara desert.

But the location at one of the larger bodies of water also resulted in many advantages at the other end of the consumption chain. The disposal of waste from larger settlements causes significant effort and space requirements that can be noticeably reduced by means of water. Consequently, as far as possible, waste was disposed of in flowing waters that transported it off. For this reason, until far into the 20th century, many rivers that flow through large cities had developed into cesspools, the smell of which was hardly endurable. Only the large-scale installation of canalization systems that flow into efficient sewage treatment plants brought this situation under control, at least in the developed areas of the world.

As the historic examples of Troy and Venice show, cities at water are often at a strategically very favorable location; they are advantageous when it comes to exchange with distant trading partners, and at the same time, for controlling wide-stretching coastal areas and trade routes. Of course, what allowed the merchant fleet to set off to distant harbors also served to bring troops up to the immediate vicinity of the destination of campaigns of conquest. Therefore, for the foundation of cities, especially coastal areas that formed natural harbors, were well

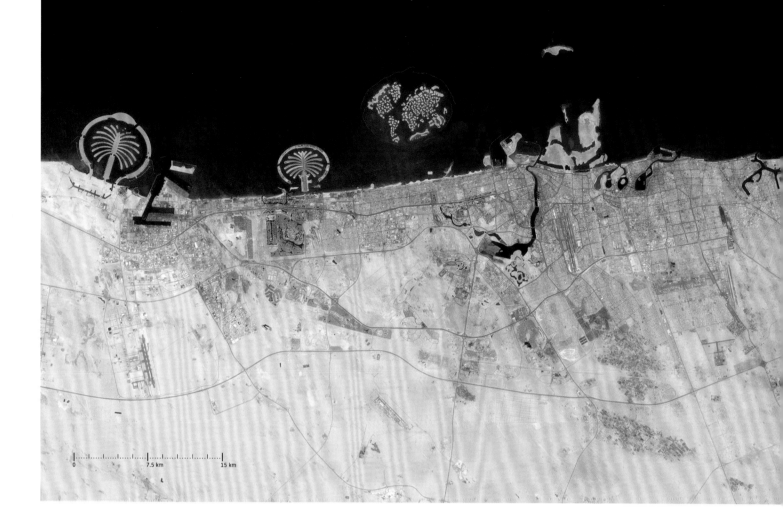

suited. Here, it was possible with relatively little effort to construct harbor facilities that were protected against the sea surf.

Often, cities were built near the mouth of navigable rivers that represented natural transportation hubs between ocean traffic and inland water transportation leading to the interior of the country. Mega cities, such as London, Shanghai or New York, are known examples that owe their development into important trading locations, not least because of this factor. Similarly, straits also represent strategically favorable locations, as is evident by cities like Istanbul and Singapore.

Water and energy sources

Being economical centers, settlements have always been challenged by a high energy demand, which was caused by continuously further developing production processes. As of a very early date, hydraulic power was deployed for the energy supply. The existence of scoop wheels driven by hydraulic power (with which it was possible to trans-

port water to a higher level) has been proven in Egypt, Mesopotamia and at the Indus, at a time period approximately 3,500 years ago. In China, the use of such technology might well date back another 1,500 years. In addition to the windmills that were added later, over thousands of years, hydraulic power remained the only source of energy with which the muscle strength of humans and their domestic animals could be replaced. Since it was only possible to use this energy locally, mills and smithies were built where the terrain and water occurrence were suitable for this purpose.

Only the invention of the steam engine in the 17th century and the utilization of electricity as of the 19th century, started to change the situation. But water power still plays an important role now for the supply of humans with electric energy.

In addition to the use of water as an energy source, its use in numerous production processes was, and still is, also an important reason for the development of settlements and cities near bodies of

↑ Artificial island worlds
25°10' N · 55°16' E (February 11, 1985 and April 13, 2013)

Enormous efforts are undertaken for the development of the area around Dubai and the artificial island worlds off the coasts of the Persian Gulf — on the one hand, because the adjoining desert imposes many restrictions, but on the other hand also because an end of the mineral oil resources is foreseeable. The fear is that deposited islands could change the currents near the shore and deteriorate the quality of the water.

0   7.5 km   15 km

water. Ancient trades, such as tanneries, dye mills or paper producers were dependent on water for their work. Often, the bodies of water were severely affected by these activities.

Disadvantages

The advantages of the location of settlements at or near water that were decisive for their foundation at this particular location are often linked to disadvantages. Those are often connected with a shortage or excess of water. Since they can endanger the existence of the affected population, for a reason, floods and droughts have always been some of the great plagues that can affect humans.

Therefore, settlements at water must be designed to withstand the consequences of extreme weather situations. This applies regardless of whether it refers to an increased flow of water due to severe rain upstream, or floods near the coast due to tropical hurricanes. At seacoasts, ebb and flow additionally come into play, which can flood wide coastal areas (especially during spring tides and in combination with strong shore side winds, during storm surges). Areas that are only slightly higher than sea level or even below sea level, such as vast areas in the Netherlands that were created as a result of land reclamation, are especially endangered. These areas must be protected from floods with comprehensive measures such as levees, and sluice and pump systems.

As recently shown in the flood in Central Europe in May and June 2013, and the typhoon Haiyan in November 2013, there are limits to preparing for extreme events. The tsunamis in Southeast Asia in 2004 and in Japan in 2011 also notably demonstrated this — no reasonable structural protection for settlements near the coast is possible against waves that are over 100 feet (30 m) high.

↑ In the riverine oasis of the Nile
15°21′ N · 32°42′ E (October 17, 2013)

The city of Khartoum was founded in 1821 where the Blue Nile and the White Nile meet in the south of old Nubia. Originally an Egyptian military post, it quickly developed into a trading base that played an important role in the slave trade, and later into the capital city of Sudan. Mainly in the Gezira Plain between the Blue Nile and the White Nile, large areas are irrigated for agricultural use.

At the birthplace of the Amazon →
3°08′ S · 60°10′ W (July 19, 2013)

As early as 1669, a fort was founded at the river bank, surrounded by rain forest, where the Rio Negro and the Rio Solimoes join to become the Amazon River. It later developed into the city of Manaus. Today, Manaus, which is located centrally and can be reached by ship over the Amazon, as well as by vehicles via the Transamazonica, is the economic center of the Amazon region.

The transformations that have been caused by climate change create particular challenges, especially in connection with water. Although the water balance in the river has been affected by global warming, it is mainly the rising sea level that directly affects the settlements and cities near the coast. The example of the island state of the Maldives shows that current developments are threatening the existence of areas that are only slightly above sea level.

Future at the water?

In recent decades, the growth of the world population together with the increase of life quality has led to a significant increase of the water demand. By now, over half of the human population lives in cities due to the increasing urbanization; scenarios indicate that this number will rise to 80 percent by the year 2050. Eight out of 10 of the Earth's mega cities with the highest population numbers are located either directly at the coast of an ocean or at a distance of less than 30 miles (50 km). As mentioned above, the concentration of areas near the water is shown particularly impressively on satellite images of Earth that were taken during satellite overpasses at night. Here, the coastlines of oceans and lakes emerge clearly in many regions of Earth, and the illuminated spots of cities appear like pearls on a string along the large rivers.

Resulting from the increasing concentration of the population in urban residential areas, the requirements for a sufficient supply with clean water surge. Cities like New York or Shanghai are long since dependent on comprehensive measures for securing the water supply that range from the massive usage of bodies of groundwater and transport from distant water reservoirs to the treatment of drinking water purification from waste water and sea water.

In this connection, it becomes more and more important to use water carefully. Waste and pollution should be avoided as much as possible to ensure that the sufficient supply for the population remains intact.

Against the backdrop of its varied options for use and its vital importance, the usage of water is one of the oldest sectors of social and political life that is ruled by laws. The varied economical and social interrelations of modern societies have led to the fact that these rules must reach beyond country borders. As shown by the example of the Nile, in which Ethiopia, Sudan, South Sudan and Egypt have a direct share, international agreements contribute to ensure the supply of the population with the essential water even in the future. Against the background of rapid urbanization, the data gathered by Earth observation satellites provides important sources of information in many areas. They play a special role where this information is not accessible in other ways due to the economical or political situation.

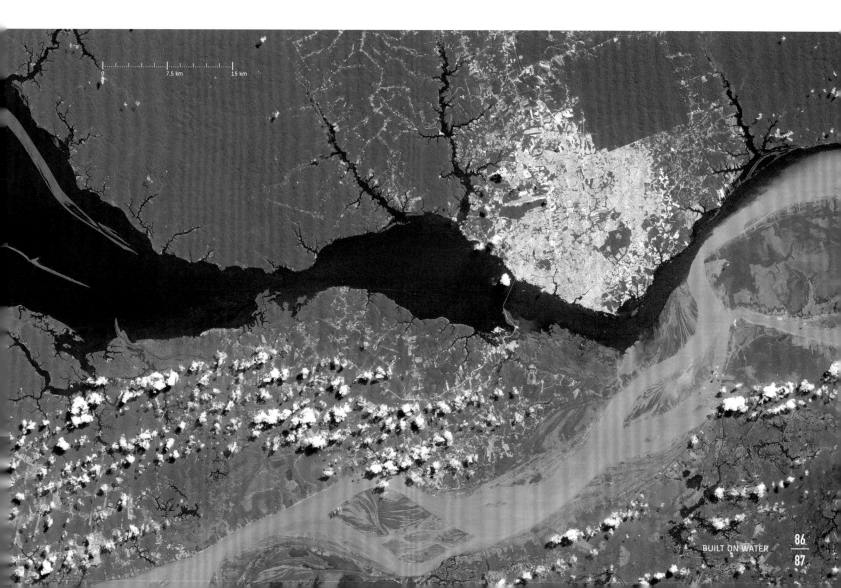

Always well fortified

55°41' N · 12°36' E (May 18, 2013)

Originally a Viking village, founded in the 10th century, Copenhagen became Denmark's capital in the 15th century and developed into an important trade and fortress city. The structure of the inner city reflects the defence fortifications that were supposed to provide safety against attacks from the sea and land as of the 17th century.

White buildings on black rock

36°25' N · 25°26' E (August 7, 2012)

Santorini, called Thera in ancient Greece, is a
volcanic island in the Aegean that has been
populated for millennia. The island received its
shape approximately 3,500 years ago, when an
explosion of the volcano left today's broken ring
structure with steep interior crater walls that are
as high as 985 feet (300 m). With its attractive
contrast between the dark volcano rock and the
white, chalked-up houses, Santorini has become
one of the most important tourist destinations in
the Mediterranean area.

SWIMMING CITIES

When the *Allure of the Seas* pulls into harbor, it is as if
an additional part of the city is berthing. Currently the
largest cruise ship, it reflects impressive dimensions: With
a length of 1,190 feet (363 m) and a width of 217 feet
(66 m), it offers space on 16 decks for a total of 5,400
passengers to enjoy a relaxing vacation. In the worldwide
increasing tourism industry, cruises are among the
strongest growing segments. The environmental impacts
of these huge floating hotels will be limited with the intro-
duction of higher standards for fuel quality and stricter
rules for routes. This also intends to prevent disasters like
the *Costa Concordia*, which had an accident off the coast
of the Italian island Giglio and took the lives of 32 people.

0   150 m   300 m

Frosty oasis

77°51' S · 166°40' E (January 31, 2012)

McMurdo Station on Ross Island at the edge of the arctic Ross Ice Shelf is occupied year-round and is a special place to live, especially during the arctic winter. The mean annual temperature is already as low as 2°F (−17 °C) and the lowest measured values so far reached almost −58°F (−50 °C). For some years, the only atomic reactor in Antarctica was used as an energy source.

From a penal colony to a metropolis
33°52' S · 151°12' E (February 2, 2011)

Far away from home, England founded a penal colony
near today's Sydney in 1788 — exactly at the same
location where Aborigines had settled 20,000 years
ago. Today, having reached a population of 4.5 million,
Sydney is the largest city and the economic center of
Australia. Well-known landmarks of the metropolis
are the Opera House, which is located at the harbor
entrance and the Harbour Bridge.

Dream destination for tourists

16°29' S · 151°42' W (March 8, 2013)

Together with Tahiti, Bora Bora is part of the French Society Islands and fits the true definition of paradise. In the middle of the Pacific Ocean, white beaches and palm trees form the basis for tourism, which represents the most important source of income for the island. Bungalows stretch out over the shallow water of the lagoon, especially along the barrier reef, which protectively surrounds the central island.

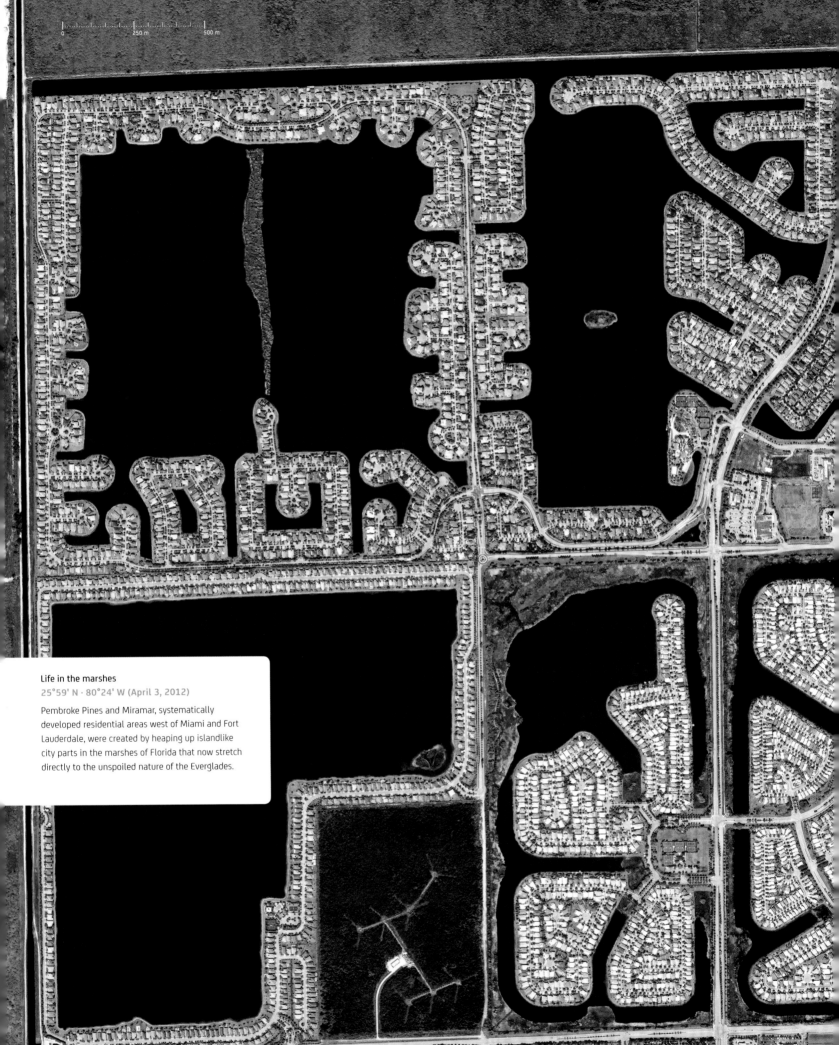

Life in the marshes
25°59' N · 80°24' W (April 3, 2012)

Pembroke Pines and Miramar, systematically developed residential areas west of Miami and Fort Lauderdale, were created by heaping up islandlike city parts in the marshes of Florida that now stretch directly to the unspoiled nature of the Everglades.

0   250 m   500 m

Self-built island kingdom

15°49' S · 69°59' W (September 24, 2009)

Lake Titicaca is located at 12,470 feet (3,800 m) above sea level on the plateau of the Altiplano in the borderland between Peru and Bolivia. With an area of almost 3,245 square miles (8,400 sq. km) it is the largest lake in South America. People of the Uros ethnic group live in the wide reed belts on self-built reed islands, some of which are occupied by more than 40 people.

0   150 m   300 m

EXPOSED TO THE ELEMENTS
Coasts as focal points for risks from natural hazards

For humans, coastal areas are very attractive. However, due to their exposed location they are also susceptible to natural hazards to a great extent; and the risk from natural disasters is on the rise. Growing investments, increasing population numbers in the coastal areas and the consequences of climate change are main contributors for the increase of the risk. To live with the risks and to reduce them requires an adequate risk awareness in all sectors of society. Coastal areas are particularly affected by natural events. Weather phenomena, such as tropical hurricanes, winter storms and storm surges rage here mainly, but earthquakes and volcano eruptions also occur frequently and are sometimes even more intensified by tsunamis.

A natural event only then turns into a natural disaster when humans or their property are hurt and damaged. So the risk mainly depends on the probability of a natural event and the presence of affected people and assets. The reason for the high risk at coasts is not only that those areas are exposed to a higher number of more severe natural events, but also the high population densities and concentration of assets.

The rise of the sea level and the increasing frequency and intensity of tropical hurricanes and floods in connection with the location of mega cities and their economical development further increase the risk to coastal regions. This can be proven impressively by the numerous disasters of recent years.

More risk by increasing population numbers near the coast

On the basis of a study from the Organization for Economic Cooperation and Development, for the 20 cities in coastal areas that will presumably be the most exposed in the year 2070, the future situation was compared to today's situation — with respect to the number of people who live in the 100-year floodplain, as well as to the tangible values that are available there.

The population of the areas at risk in these 20 cities will be approximately 113 million people by 2070; this is almost five times as many as today. With the exception of Miami and New York-Newark, all cities on this list are located in Asia and Africa. Some of them, like Dhaka and Chittagong in Bangladesh, are expected to grow by more than tenfold. In all 20 cities, current assets in the amount of $2.2 trillion are at risk. In the year 2070, this value will amount to approximately $27 trillion, an increase by twelvefold. Accordingly, the number of people and assets in flood-prone areas will significantly increase.

Effects on humans

With the global temperature rise, the sea level also rises. Many small islands and large coastal sections in some poor countries will sink in water by the end of the century. Other regions will be able to prevent permanent flooding, but will have to deal with more frequent flooding or invest enormous amounts of money to maintain the current protection standard.

The heights of waves and storm surges will increase as a result of more frequent and stronger storms in many parts of the oceans. This will not only increase the threat to the coasts, but also to ship traffic and offshore platforms. The North Atlantic and western Pacific, where the coasts with the highest asset concentrations in the United States, Europe, Japan and China are located, will experience the most severe changes.

Precautions against natural hazards

The most efficient type of risk reduction is to not move to a location at risk. However, it is difficult to prevent people from moving into areas at risk if they hope for a better life in these areas. And trying to convince them to move away again is almost impossible.

Coastal cities must therefore be protected with technical means against storm surges and other risks that originate from the sea. The same applies to structures that are naturally exposed to these risks — harbors, oil platforms in the open sea and wind farms.

A lot has already been achieved. In the years 1970 and 1991, 300,000 and 139,000 people still respectively died during storm surges. Over the past 25 years, thousands of solid elevated emergency shelters have been constructed, to which people can escape during flooding of the area. During some storm surges in recent years (that were comparable to the two mentioned above), fewer than 5,000 people died each time. In contrast, in May 2008, 140,000 people in Myanmar fell victim to the cyclone Nargiswhich, which hit the completely unprepared delta of the Irrawaddy.

Improved building codes, levees, monitoring and communication networks, and the availability of satellites offer new possibilities. At the same time, our growing understanding of physical processes and their modeling leads to continuously improved early-warning and alarm options. However, even with the best technical measures, absolute safety will never be achieved.

Spreading the risk

The flood risk must be spread across several shoulders: the state, the affected citizens and companies, as well as the financial sector, especially the insurance industry. Disaster precautions can only be effective if all these parties cooperate in a sensible partnership.

In this, the main task of the state is to minimize the risks for society as a whole by installing early-warning systems, building levees, identifying flood areas, specifying a legal framework for the utilization of land in areas at risk and preparing emergency plans that also include programs for rebuilding.

Directly affected people can also contribute a lot to the mitigation of damages. In this, the risk awareness is decisive. Everybody who wants to construct a building should be informed and obligated to build reasonably. This individual would have to check him/herself to which extent his/her property is at risk and accordingly take the required measures for a potential emergency. This also includes financial precaution measures with respect to losses to be expected, for example by taking out insurance.

Unavoidable developments

We must admit that the concentration process in coastal regions will continue in the

foreseeable future and also that climate change is a fact. We cannot stop this development or even reverse it — at best we can slow it down. The rise of the sea level will result in some coastal areas currently used by people having to be abandoned due to flooding. This development will create climate refugees, especially in poorer countries. The risk reduction at the coasts must be accomplished by adjusting to the new conditions. In the process, advanced engineering technologies, and also rigorous emergency procedures, become more and more important. Due to their size, island states and delta regions can hardly be protected with technical structures.

For all areas, the quality of early warnings, alarms and evacuation measures must be further optimized. The best measures for early warnings are useless, however, if they are not implemented into correct actions by people. For this purpose, understandable information, good education and mainly the corresponding awareness-raising is necessary.

Dr.-Ing. Wolfgang Kron is responsible for everything to do with "water as a natural hazard" in the Geo Risks Research department of Munich Re, the world's largest reinsurance company. The civil hydraulics engineer's specialty is flooding risk and its reduction at coasts and rivers, but also far away from them.

"Coastal regions are the risk focal points of Earth. This is where nature rages and this is where humans and their assets are massed."

Already in the medieval times, Venice was the center of an expanded trade kingdom that spanned the entire Mediterranean area. The basis for this position was the unique location of more than 100 islands in a lagoon at the northern Adriatic Sea. A dense network of canals runs through the world-famous historic city center of Venice, which is exposed to regular floods at "acqua alta" and dense streams of tourists.

PROTECTION OF HISTORIC HERITAGE

For cities that are located directly at the water, any change of the sea level often has grave consequences. Already a slight rise — whether it is because the subsoil sags or because the water actually rises — entails considerably higher risks concerning storm surges, for example. In Venice, both factors exist and caused that the sea level in the city was 9 inches (23 cm) higher at the end of the 20th century than it was at the beginning of the same century. This resulted in the number of floods noticeably increasing, as well as the extent of the floods. In order to avert the threat from the city, the construction of MOSE began in 2003. After the completion of this project, when required, mobile flood protection gates were able to be activated; these gates close the three openings of the Lagoon of Venice to the Adriatic Sea and are intended to protect the lagoon and Venice itself against floods from the Adriatic Sea.

0   300 m   600 m

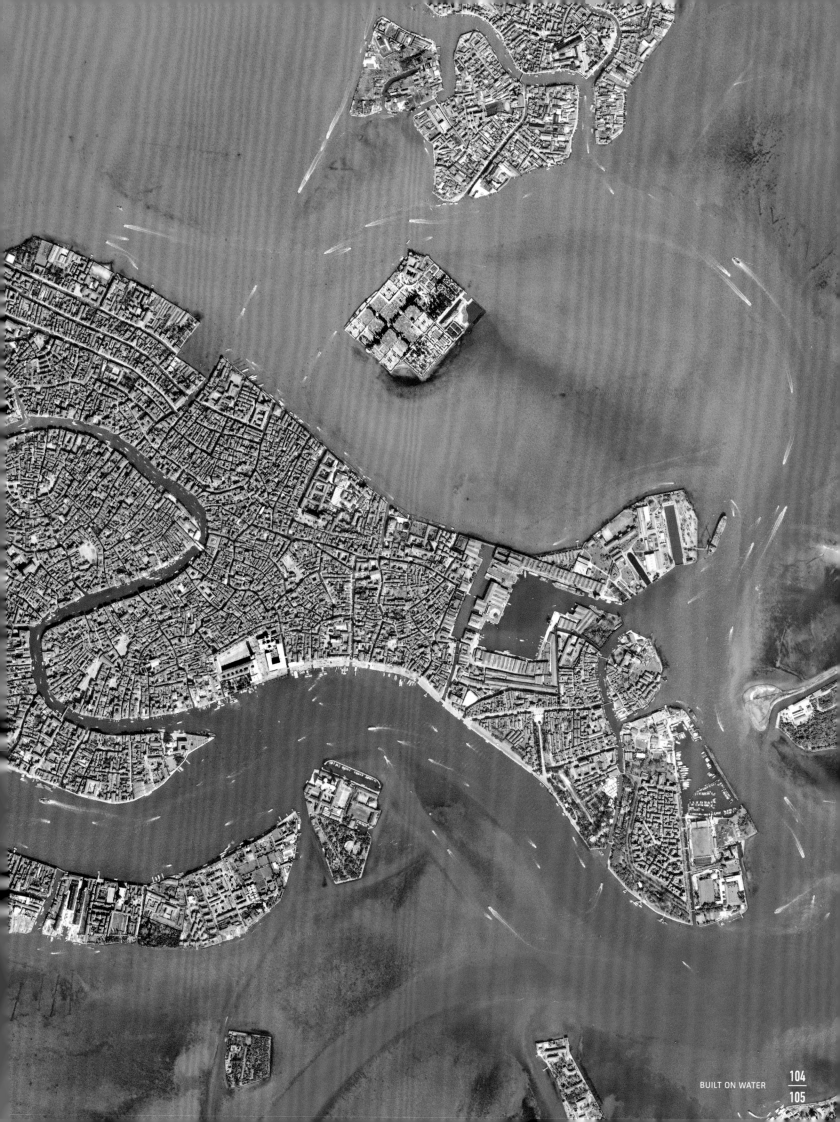

Food from the sea
14°50' N · 120°44' E (December 26, 2009)

The city of Hagonoy is located on Luzon, the main island of the Philippines. The city that mainly lives from fishery sprawls along a canal that stretches through the coastal plain of the Manila Bay and thus provides good access to water. Aquaculture basins for the farming of shrimp, mussels and fish cover almost three-quarters of the area.

0　　　　200 m　　　　400 m

Metropolis in the cold
64°11' N · 51°43' W (June 26, 2010)

Greenland's capital city Nuuks is located near the
Northern Arctic Circle at the bay of the Labrador Sea
and is one of the largest arctic cities, with a population of 14,000. At a mean annual temperature of
29°F (−1.4 °C), living conditions are hard. Regardless
of that, traces of settlements go back more than
3,000 years.

0   200 m   400 m

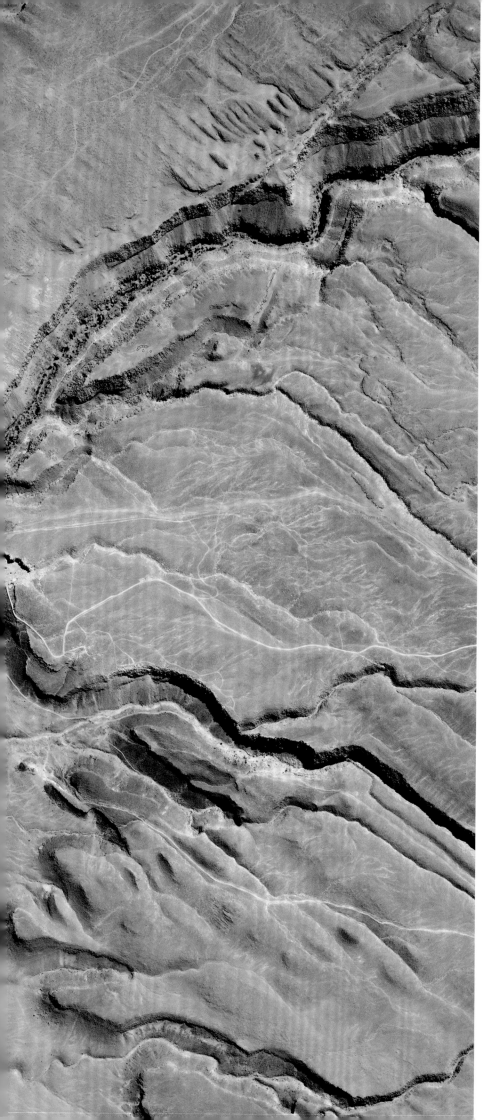

Water for the Salar
23°11' S · 68°00' W (September 21, 2010)

The village of Toconao is located at the northeastern edge of the salt lake Salar de Atacama in the Chilean region of Antofagasta. The village is located at about 8,200 feet (2,500 m) above sea level at the end of one of the few watercourses that supply the Salar with water, at least from time to time, in these extremely dry surroundings. The population of Toconao, some of whom work in the salt production at the Salar de Atacama, are also completely dependent on this watercourse.

LIFE IN THE DESERT

Life in the desert has always been associated with deprivation and the necessity to come to terms with the special conditions of this habitat. In these surroundings, water plays an even more direct vital role than in other places when it comes to the question of survival. While in dry regions of steppes, at least extensive livestock breeding by nomads is possible, people in desert regions are dependent on oases that have a sufficient water supply. Often, there are very special reasons that make living at these locations worthwhile. One of these reasons may be natural resources that can be exploited nearby, but primarily, trade is the most important factor. Over centuries, trading bases in the Sahara Desert prospered — for example the city Timbuktu in Mali — with their water supplies being vital for large caravans crossing the desert.

Spots in the "Green Hell"

1°29' N · 68°14' W (August 22, 2003)

The Rio Içana is one of the smaller tributary rivers in the rain forest region of the Amazon Basin. In the northwest of Brazil, the river runs through the borderland to Columbia and flows into the Rio Negro that joins the Rio Solimoes at Manaus to become the Amazon River. The inhabitants fish in the river and practice shifting agriculture for which smaller forest areas are cleared to make space for the cultivation of agricultural crops.

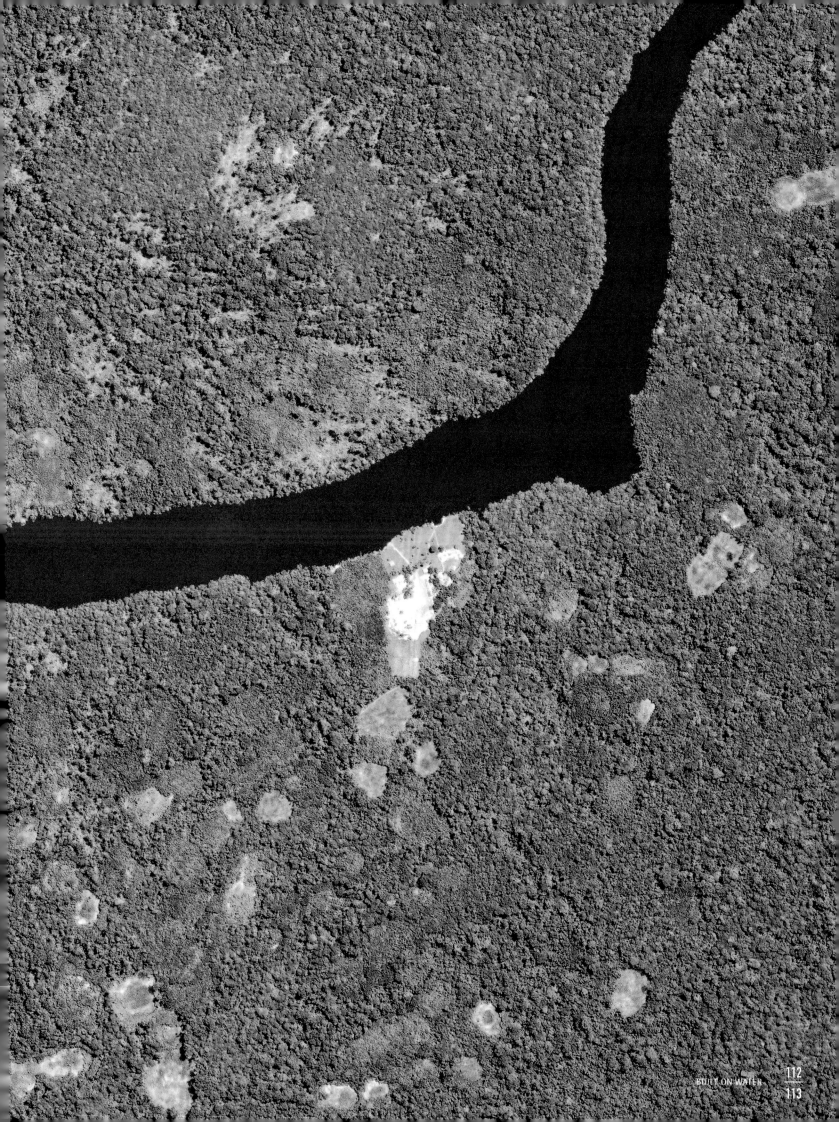

Human-made recreation

27°23' N · 33°40' E (January 5, 2013)

Located north of Hurghada at the entrance to the Red Sea, Al-Gouna is part of the Egyptian Red Sea Riviera. The vacation resort was completely developed on the drawing board, with no need to consider existing structures. Golf courses and hotels are grouped around a system of canals that allow all guests direct access to the water.

0   300 m   600 m

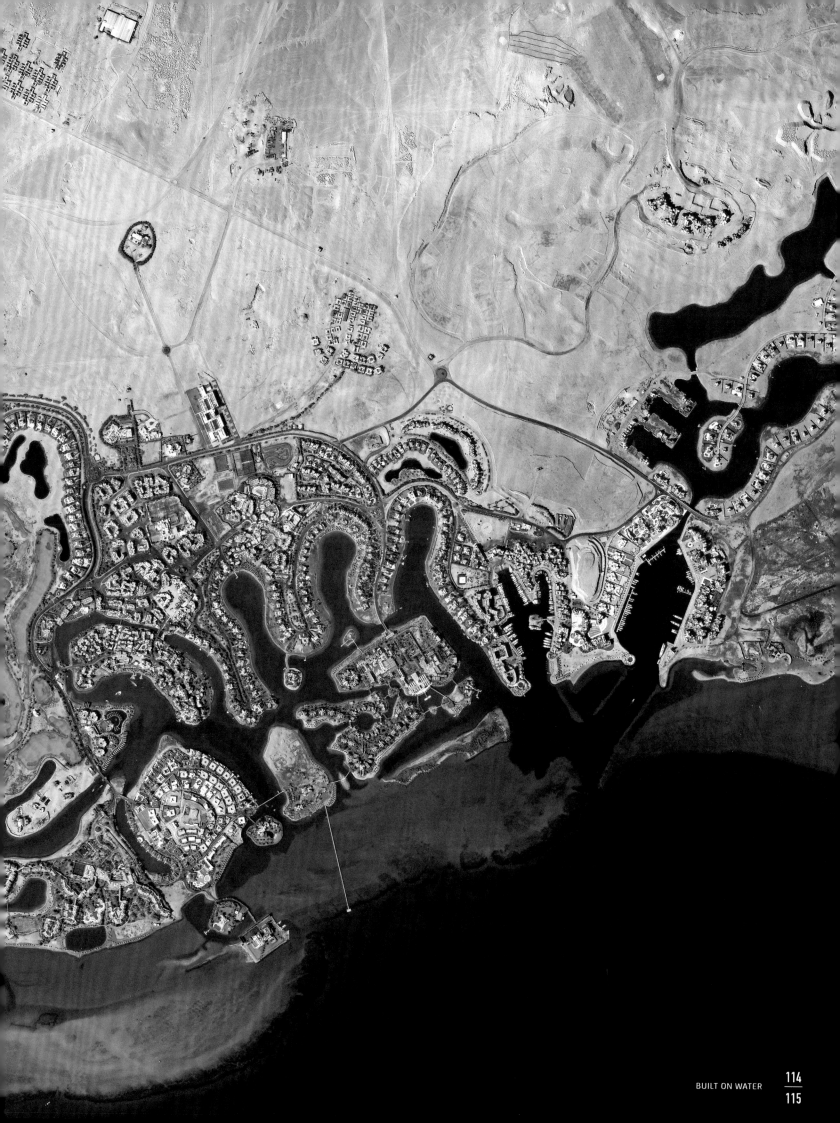

0   200 m   400 m

The Gate to the New World

40°43' N · 74°00' W (October 7, 2011)

With the characteristic skyline of the central urban district of Manhattan, New York City is the prototype of a modern metropolis. At the southern tip of Manhattan you will find piers from which ferries connect the island with New Jersey and other neighboring cities.

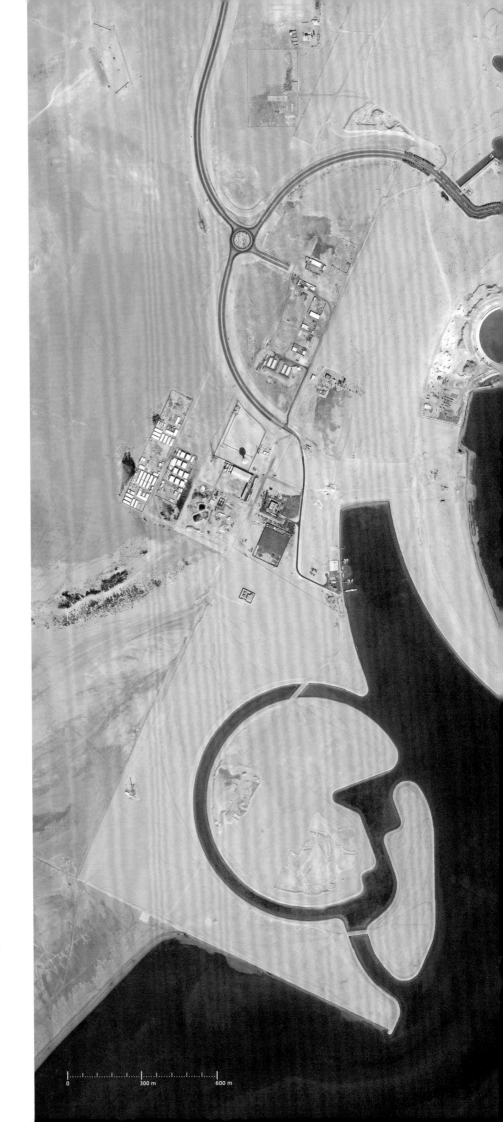

Beautifully strung

25°50' N · 50°36' E (November 13, 2013)

By 2015, 15 artificial islands that are connected
with bridges shall form Durrat al-Bahrain, which
is reminiscent of a necklace with fish-shaped pen-
dants, at the southern tip of the island state
of Bahrain. Similar to the projects "The Pearl" and
"The World" off Dubai, surfaces surrounded by the
sea shall be created. It is intended to create
a pleasant climate for living in the hot desert
climate of the region.

ARTIFICIAL WORLDS

The artificially built island worlds of the countries in the
Persian Gulf, which became rich from oil, have caused
controversy time and time again over the past few years.
They are counted among the most elaborate individual
projects with which landscapes were changed. In addition,
probably for marketing reasons, their structures are aimed
at visibility from space and thus prompt discussions easier
than other environmental intrusions that might be even
more severe. For the construction of the islands that are
several miles in size, hundreds of millions of tons of sand
and rock are moved; for the stabilization of the shore
areas, wave breakers made of solid rock blocks must be
created. After their completion, the islands also serve as
platforms for residential and commercial buildings; however, the main objective is the construction of hotel resorts
in order to boost tourism in the region.

0   300 m   600 m

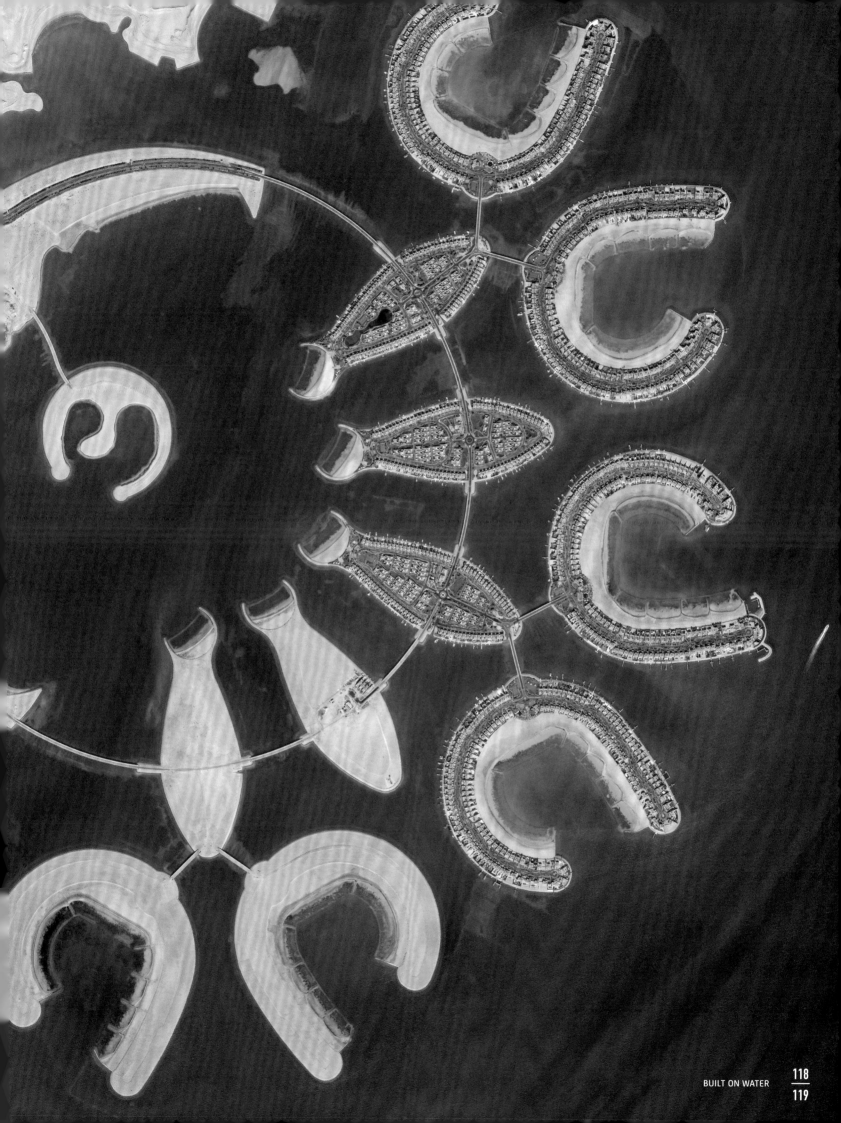

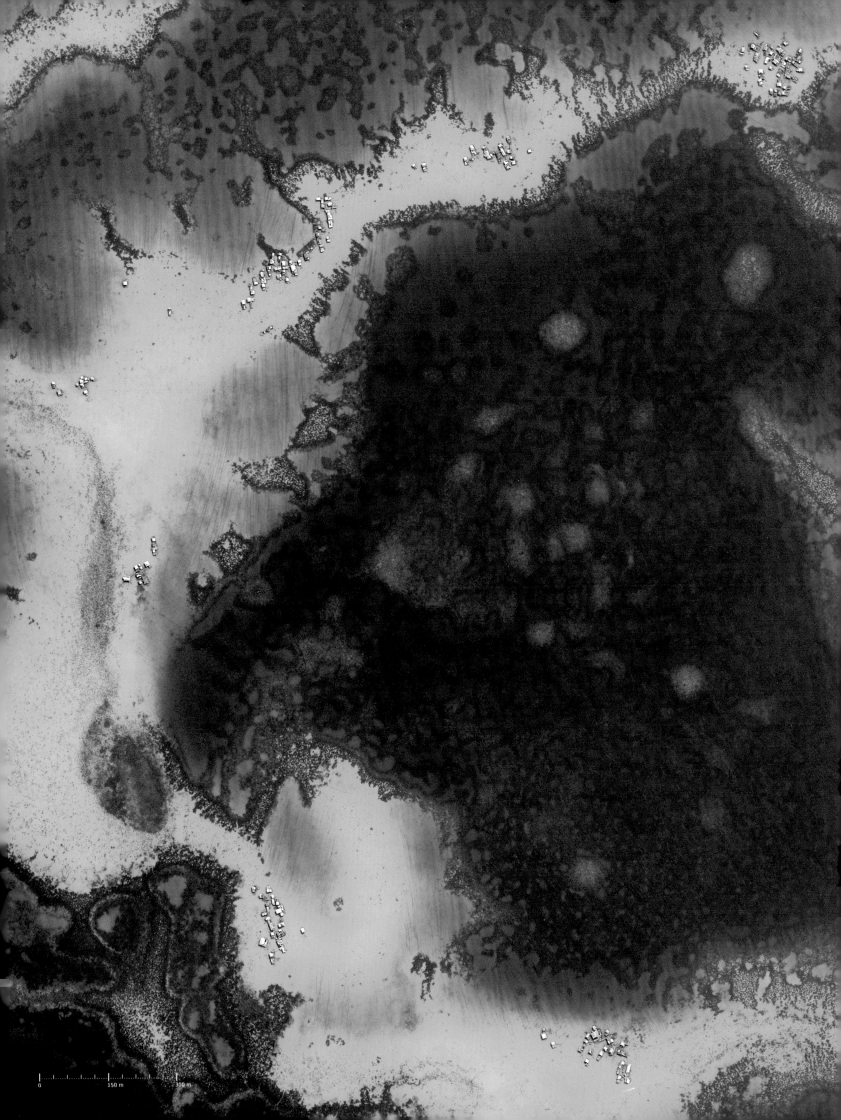

The stilt houses of Borneo

4°37' N · 118°39' E (March 16, 2013)

The Tun Sakaran Marine Park, which consists of coral reefs and small islands, is located a few miles off the northeastern coast of Borneo and features a multifaceted underwater world. The Bajau Laut who used to live on boats as sea nomads, live in the west of this region. Today, however, most of them have settled, often by erecting their villages on stilts in the sea as shown in this image.

0   150 m   300 m

Vacation between metropolis and the sea
22°59' S · 43°11' W (January 3, 2010)

Although Copacabana refers to an entire district of the
Brazilian city of Rio de Janeiro, the name is generally
associated with the 2.5-mile (4 km) long sand beach
between the city and the surf of the Atlantic Ocean.
Along the beach, one hotel is followed by another.
Together with the beach of Ipanema, the Copacabana
is one of the pillars of tourism of the city.

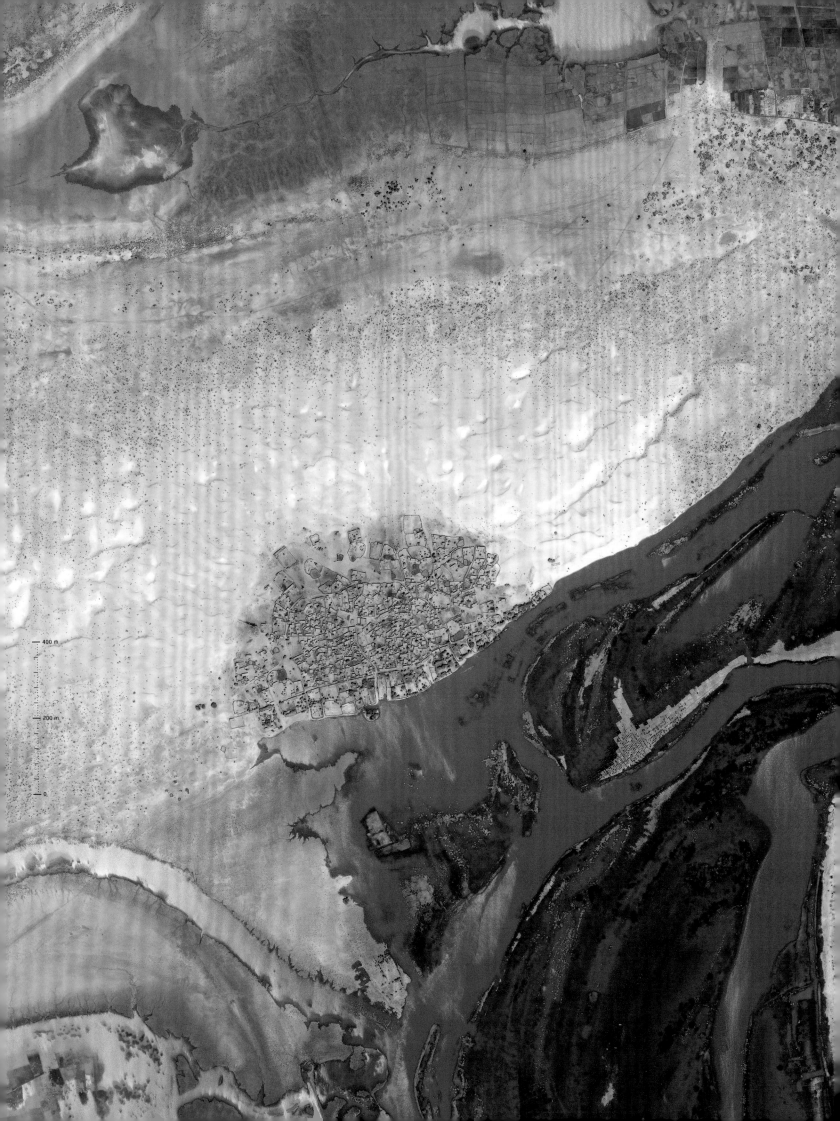

400 m

200 m

0

Niger, the lifeline

16°12' N · 3°18' W (October 17, 2007)

The small town of Minéssengué is located south of Timbuktu, in the exit region of the Niger inland delta, at one of the numerous river arms of the Niger River. Due to its proximity to the Sahara, the region is dry and dependent on the water supply from the Niger, which enables the population to practice irrigation cultivation and fishing.

Unperturbed by tourism

20°54' N · 107°01' E (February 25, 2010)

Four swimming villages exist in the picturesque surroundings of the Vietnamese Halong Bay, where fishermen live in the middle of numerous tourist boats. More than 1,600 people have direct access to their food, which consists of fish, mussels and aqua-culture products.

0   150 m   300 m

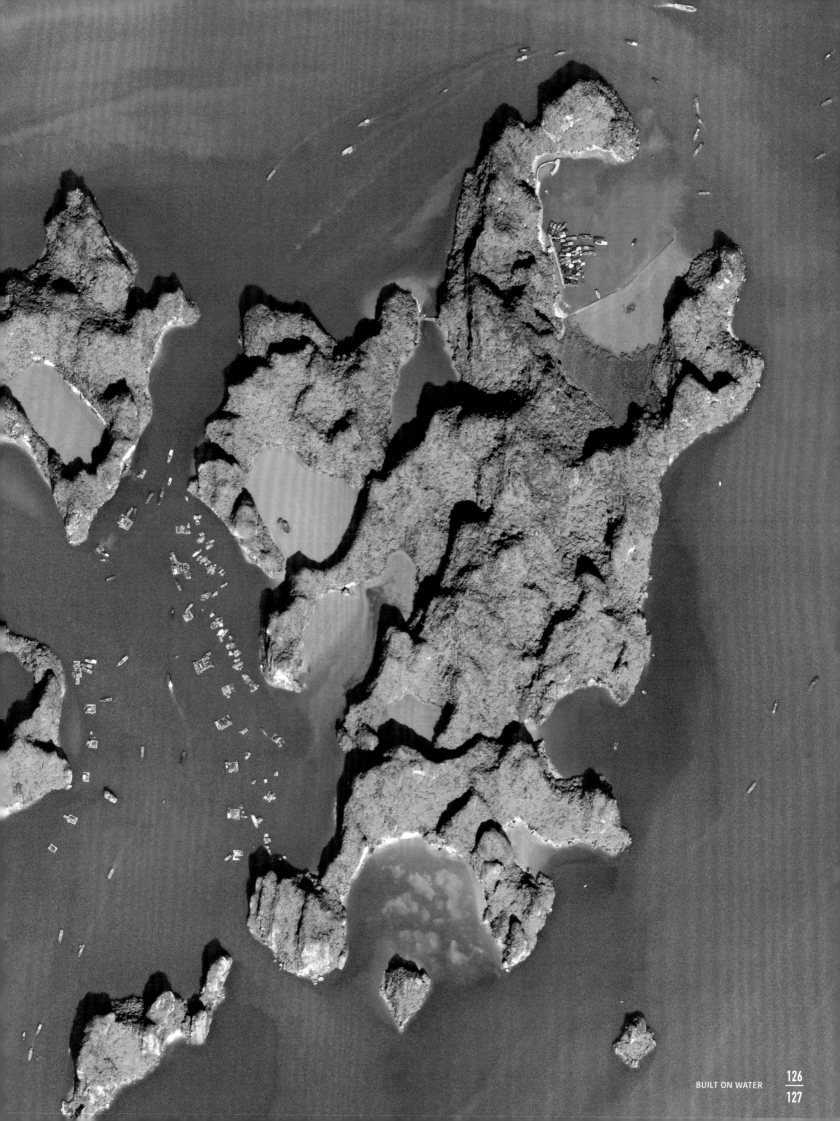

Traces of rapid development
31°14' N · 121°30' E (May 11, 2009)

Since the economic opening of China in the 1990s, as one of the largest mega cities in the world, Shanghai has undergone a rapid development process. The sky-scrapers in the business district of Pudong are in clear contrast to the remainder of the residential districts that were originally built with one- and two-story houses on the other side of the Huangpo River and that will eventually vanish due to modernization.

150 m   300 m

Water for millions

31°14' N · 121°38' E (August 29, 2013)

More than 24 million people who must be supplied with drinking water live in the area of Shanghai. In addition to the groundwater, mainly the rivers Huangpo and Yangtze, which now cover 70 percent of the water demand, are used. In 2010, the new Qingcaosha Reservoir with a size of 27 square miles (70 sq. km) at the northern edge of the island Changzing Dao, which is located off Shanghai in the mouth of the Yangtze River, was taken into operation. The reservoir, which clearly stands out in the satellite image from the brown waters of the river, can cover the water demand of the city for 68 days.

COMPREHENSIVE PROBLEM SOLUTIONS

For the growing mega cities of the world, water becomes more and more significant. This concerns the supply of clean water, as well as the disposal or reprocessing treatment of polluted water. It therefore becomes increasingly important to reach an integrated water resource management (IWRM) that includes all levels of water utilization. In the course of this, technical and economical questions must be considered and solved; political, social and ecological aspects also play an important role. Therefore, affected parts of the population, but also institutions and organizations, must be included in the important decision processes from the beginning. Training, awareness-raising and information are also part of this comprehensive approach.

City at the Golden Horn

41°03' N · 29°11' E (July 30, 2013)

Since 1950, the population of Istanbul has almost quintupled to 14 million people. One of the reasons responsible for this rapid development is the strategic location at the Bosporus, which connects the Black Sea, and thus the Eastern European and Central Asian regions, with the Mediterranean Sea. Every year, 50,000 ships travel this waterway. Another canal is now planned to relieve some strain.

Casino city in the desert

36°09' N · 114°43' W (April 16, 2014)

The example of the desert city Las Vegas illustrates the role of water for settlements in a special way. The city developed mainly due to gambling and the associated tourism. Today, Las Vegas has the highest per capita drinking water demand of the United States with 219 gallons (830 L) and is completely dependent on the supply from Lake Mead, which is located east of the city.

WHEN THE FLOOD ARRIVES

Floods are a manmade problem

The increasing number of floods is not a result of climate change. Rather, the responsibility lies with the senseless agricultural and settlement policies that take away more and more space from our rivers and prevent their natural development.

In Passau, people are used to living with floods. Since its foundation in Roman times, the city, where the Danube River is joined by the rivers Inn and Ilz, holds up against the floods. And it will brave them in the future. Or will it not? The flood in early June of 2013 shook the confidence of the people of Passau. During the most severe flood of the last 500 years, the water levels rose above the historic records. Damages were never that high before. Now many inhabitants of the Bavarian "City of Three Rivers" look ahead anxiously.

Floods are no longer exceptions. They happen almost every year, cause extensive damages, traumatize the affected population and claim human lives. Here, the number is just a few, but in the tropics and subtropics, thousands of lives are lost. Almost every day, we hear news about disasters. Their presence in the media desensitizes us. For a few days, floods are raging and there is great concern. Then "Father Rhine" and "the beautiful blue Danube" flow gently again and passengers on the excursion boats enjoy a deceiving idyll.

Appearing and vanishing

Floods are a part of the nature of rivers, the same as low water. Floods provide fresh nutrients to meadows, create new structures with their destructions and thus generate the high natural biodiversity that characterizes river meadows. The flow of water of a river never stays the same; it changes over the course of a year and over many years. Floods originate in trenches and small creeks. Continuing severe rain triggers them. The larger the drainage area of rivers, the more severe floods can become and the steeper the slope of the feeders, the faster the flood arrives.

People settling at rivers always knew that. Over centuries, villages and cities were built where the floods could not reach them. The meadows — the name means "water grassland" — remained free from settlement and fields; they were used as pastures instead.

Only at a few locations, one had good reason to build directly at the water (for example, when harbors or customs offices were needed). Unavoidable flood damages balanced the advantages of the location directly by the river. The lower floors of the buildings that were exposed to floods here were neither lived in nor used for stockpiling. For centuries the interplay between gains and losses worked — the water often took a lot, but in the end it gave even more.

About the disappearance of the alluvial forests

At the beginning of the 19th century, the situation changed. Rivers were regulated by dams and levees. However, water flows a lot faster through a canalized bed than in its natural state. This results in the deepening of river beds. If previously smaller floods were common, now there were none. The water did not reach the meadows anymore, and since the groundwater levels fell in conjunction with the deepening of the river beds, the meadows dried up. Straightening of rivers in favor of the shipping sector had further, quite desirable consequences. Now, it was possible to use the meadows better for agricultural purposes. Where before, pastoral farming was practiced, arable farming was now possible. The produced gains as a result of this compensated for the occasional losses due to severe floods. Higher dams kept even these floods away from the cultivated land. Since the low-lying land was now flood-free, it was possible to also build residential buildings and industrial plants. The cities moved closer to the rivers. Parts of towns that were previously separated and could only be reached with effort via ferries or footbridges, grew together thanks to solid bridges. Already in the early 20th century, rivers were permanently integrated in the city landscapes almost everywhere.

This worked fine for about half a century. It seemed as if the power of rivers was tamed. Dam lakes provided electricity and held back the floods. Disastrous floods became so rare that there was no question about the land reclamation at the rivers. That this caused the alluvial forests to shrink and disappear apart from puny remainders, at best disturbed nostalgics and some nature lovers who mourned the vanishing of the rich nature.

The really big turning point in hydrological balance came after World War II, when water management changed the entire drainage area of rivers with land consolidation and amelioration measures. Even small rivers and creeks were straightened; wetlands and marsh districts were drained. Billions of dollars in tax money was poured into the development and remodeling of an agriculture that soon did not know how to handle its own overproduction. Mountains of grain, lakes of milk and masses of meat were produced that could no longer be sold. Increasingly, the subsidized EC and EU agrarian system that was (mis-)guided by the principles of planned economy became independent.

Intrusions in the hydrological balance and their consequences

Politicians no longer dare to undertake urgently needed corrections. Large-scale agricultural technology results in increasingly frequent and more severe floods. The floods rise because precipitation is diverted faster and faster via the straightened creeks and from the drained areas. Corn growing, which has meanwhile become dominant in Germany with over 5 million acres (2 million hectares) of cropland, increases the rapid drainage of water from these areas even further. This results in consequential damages due to dirty water that is polluted with agricultural chemicals, because the corn does not yet cover the ground in early summer when severe floods most often occur. Since they were robbed of their natural flooding areas, the considerably narrowed rivers rise more severely than ever before — even if the precipitation amounts are not particularly high.

So one very significant part of floods and the damages that they cause is a consequence of human actions. They are the consequence of deep intrusions into the hydrological balance of landscapes in favor of a form of agriculture that is subsidized by society at an extremely high level, without accepting the necessity of dammed lakes and polders in return. Society pays for this undesirable development triple-fold: First for the agricultural subsidies, then for the flood damages and finally even for crop shortfalls that were caused by the unleashed floods in the agricultural areas close to the rivers. But for some building damages, the municipality is the actual party responsible. They acquired the flood-prone areas at a favorable price and developed them into building areas, without considering the increasing number of floods — there is a reason why most of the buildings that have been damaged by floods are recently built buildings. The only alternative would be to reconstruct the river beds to a near-natural condition and provide more room for them again.

Munich proved that such a renaturation is possible. For over 10 years, the Isar River has flowed into the city area in the form of a wild river again. And voilà: The flood of 2013 did not cause any major damages in the Bavarian capital. Rather, the renaturation measures created a new recreational area and increase in the quality of life for the residents of Munich. The future requires more nature at the rivers. It will provide more safety against floods than total development.

Professor Dr. Josef H. Reichholf, born in 1945, is a zoologist, evolution biologist and ecologist. He has been working at the Bavarian State Collection of Zoology ("Zoologische Staatssammlung") in Munich for over 35 years. In addition, he taught at the University of Technology and Ludwig Maximilian University of Munich. Reichholf is member of the commission for ecology at the Bavarian Academy of Sciences and Humanities ("Bayerische Akademie der Wissenschaften") and is author of numerous scientific and popular scientific publications. He is one of the most important natural scientists of Germany.

"Since they were robbed of their natural flooding areas, the considerably narrowed rivers rise more severely than ever before."

Agriculture from the chess board
32°12' N · 121°17' E (April 15, 2009)

North of the Chinese mega city Shanghai, a flat coastal plain spreads out, the fertile soil of which has always been utilized intensively. The tightly interwoven rectangular networks of canals and streets are lined with numerous small farming operations, which (in addition to their self-supply) form the backbone of food production for the entire region.

POLITICAL SETTLEMENT STRUCTURES

What are the differences in the settlement structures in different parts of the world? In addition to the geographical conditions, which have an effect on the water supply for example, the structure of landscapes, their utilization and the settlement forms, always also reflect the social and political conditions of humans. Historic developments, property situations and the extent of individual planning as opposed to centralized planning, influence the appearance especially. Whereas in the dissected old settlement areas of Europe severely irregular structures are prevalent, in the wide plains of the American Midwest, as well as in the coastal plains of eastern China, you will find clearly structured grid patterns. This applies to the structure of the agricultural areas, as well as the actual settlements.

Safer on the lake

6°28' N · 2°24' E (January 9, 2013)

Safety was always a reason for living on water, as is the case in the small town of Ganvie in the West African Benin. During the peak of the slave trade in the 16th and 17th centuries, the residents along Nokoué Lake began to construct their houses on stilts in the lake because it was forbidden for many slave hunters, due to religious reasons, to follow them into the water.

0   200 m   400 m

The heart of the empire
51°31' N · 0°05' W (October 22, 2011)

The settlement that had already been founded during the Bronze Age at a fording above the Thames has developed to the center of an empire over the past centuries. London is still Europe's most important financial center. New skyscrapers that meet the demand for office space stand next to buildings that are steeped in history, such as the Tower or St. Paul's Cathedral.

400 m

200 m

0

Priceless asset
THE ELEMENT OF LIFE

FOOD AND SALT FOR MILLIONS: Canals run through the alluvial plains near Bangkok. In addition to rice fields, this is where you will find mainly aquaculture basins for the farming of shrimp and evaporation fields for the production of salt (13°31' N · 100°28' O; January 31, 2010).

WATER — KEY ELEMENT FOR LIFE

When we talk about water, the key element for life, our mind quickly captures the image of a person, dying of thirst and being mocked by mirages, who drags himself to an oasis in the desert with his last bit of energy. This image also illustrates very directly the importance of water as a key element for life without which a human can only survive a few days. However, the percentage of water that is used as drinking water is very low in comparison to the percentage of the total water consumption.

Every person uses large quantities of water, although we are usually just aware of a small portion of this consumption. In Central Europe, a daily per-capita consumption of approximately half a gallon (1.5 l) of water is directly consumed through food and drinks, compared with a household consumption of approximately 31 gallons (120 l) per person that is mainly required for other purposes such as personal hygiene, household and wastewater disposal. If you include agri-

culture and industry into this calculation, the total per capita water consumption increases dramatically. In accordance with climate and economical conditions, the water consumption is distributed very unevenly globally — from less than 13 gallons (50 l) per day per person in countries like Rwanda and Uganda, and 343 gallons (1,300 l) in Germany, the water consumption ranges to a value of 1,188 gallons (4,500 l) per day and person in the United States.

However, it is misleading to talk about consumption of water, because in most cases it is not consumed in the stricter sense of the meaning, but rather reintroduced into the larger cycle after being used. However, most of the time, it is no longer possible to use it as drinking water after its use. Water that is used for irrigation evaporates into the atmosphere if it is not stored by the cultivated plants. Water that is used in households flows into the sewage system in the form of polluted wastewater or drains away in the subsoil.

In connection with the general discussion relating to sustainability and the terms that are often used in this context, like ecological footprint and CO_2 footprint, with which the impact of humans on their environment is intended to be expressed in numbers, the term "water footprint" was developed around the turn of the millennium. This

"Water, Water!" Unknown traveler

← **Salt from the Dead Sea**
31°10' N · 35°36' E (March 8, 2002)

Like fish bones, the dams in the evaporation basins run through the southern part of the Dead Sea, which is divided into an Israeli and a Jordanian part by the 1,640-foot (500 m) wide main dam. They form the basis for the production of salt. While the water level in the northern part of the Dead Sea has fallen by approximately 100 feet (30 m) over the last century, the level in the southern basin is rising.

term is based on the concept of virtual water that includes water that is used for the production of all other utilized products and their pre-stages, in addition to directly consumed water.

Drinking water

In the previous chapter, we gave many examples as to how the question regarding the availability of water has determined the choice of the location for founding settlements. The availability of drinking water in sufficient quantity and quality, preferably provided by existing sources like clear streams and rivers, has played an important role.

After settlements grew, it often became important to supplement the sources of drinking water that were found in the near vicinity once the quantity of water was no longer sufficient, or to replace them when the water quality was affected too severely by increasing usage. This was shown especially at rivers, since they were used for the disposal of solid and liquid waste at the same time. Even today, in many regions of the world, untreated sewage is discharged into rivers, where it causes huge hygienic problems in areas further downstream and thus provides ideal conditions for the spread of known diseases like typhus, cholera and dysentery.

To make the water fit for human use and consumption, one can simply boil the water. If there is sufficient heating material avail-

able, then this can be carried out for smaller quantities like directly consumed water without any problems, and also for water that is required for washing; this procedure takes a lot of effort. Not least for this reason, other sources that ensure the water supply over the long term with a one-time high effort have been used from very early on.

Groundwater bodies are tapped into by means of wells to provide water during long dry spells; this water has significantly fewer problems with the quality than the surface water due to the filtering effect of the ground. The oldest confirmed wells have been dug or drilled in Cyprus and Israel over 10,000 years ago, but even today, wells play a central part for the water supply.

With technology to drinking water

When the springs directly around the settlements dried up or were no longer sufficient for the water supply due to the population growth, the first catchments of springs and water pipes were built. Real masterpieces in this respect were developed by the people in the Arabic-Persian region where the system of qanats (which has proven useful over thousands of years) is often still used today. This system belongs to the oldest traces of humans that are visible in satellite images. By means of qanats, water is diverted through tunnel systems to the destinations from springs at mountainsides, which are often many

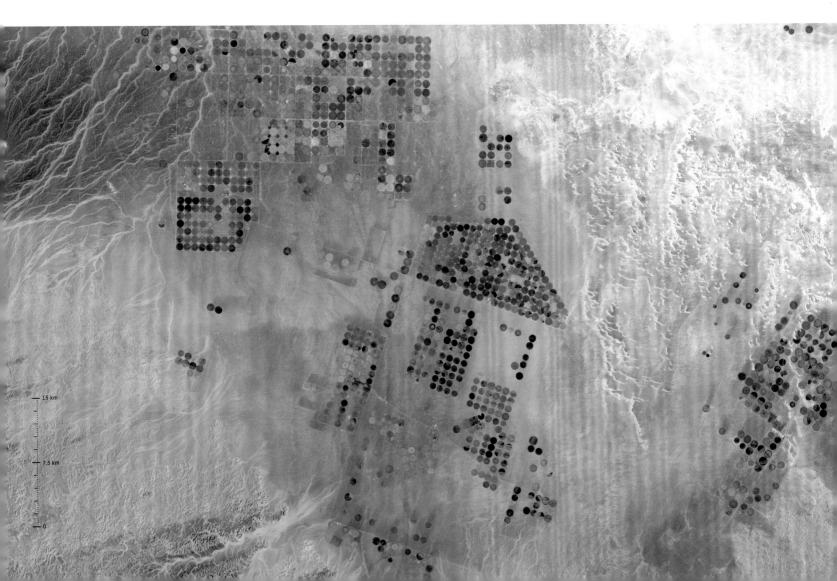

15 km

7.5 km

0

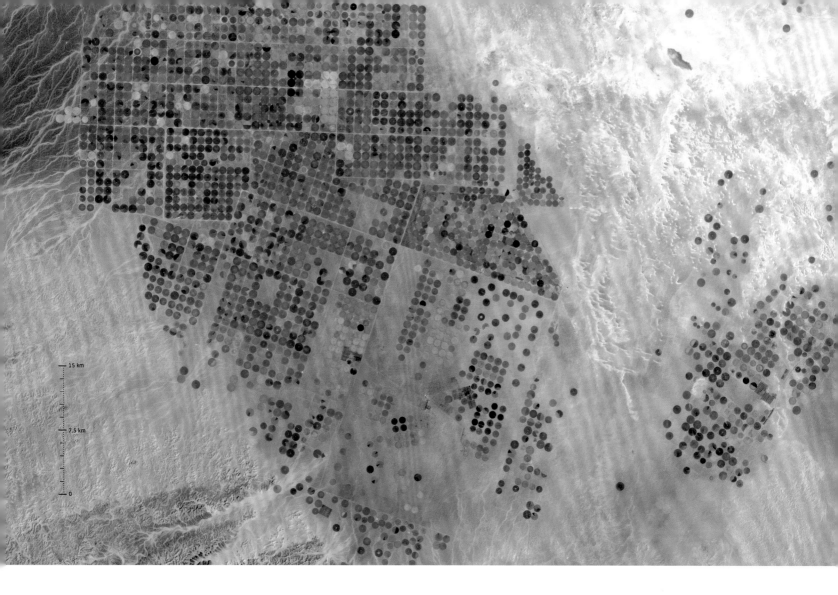

15 km

7.5 km

0

miles away. At the surface of the Earth, the qanats' characteristic chains of craterlike holes, which mark the entrances to vertical maintenance shafts at intervals of a few feet, can be identified very easily. Another technical masterpiece are the aqueducts in Rome, a city that grew into a mega city in ancient times. Up to 11 water pipes, some of which negotiated mountains and valleys via pressure pipe lines and complex bridge constructions, transported water of high quality from distances of almost 60 miles (100 km) to the center of the Eternal City, where water was required for numerous thermal baths, fountains and gardens.

Wells and long-distance water lines, including the sophisticated water line and storage systems that are required in modern large cities, represent important components of the water supply even in our time.

However, these measures are often not sufficient to meet the water demand of the highly increased population. In addition to the grown demand, the quality of the groundwater in areas intensively used for agriculture is often no longer sufficient for the use as drinking water, due to overfertilization and the use of pesticides. Therefore, water in densely populated areas is additionally treated by mechanical, biological or chemical purification to the extent that it can be used again after having been mixed with fresh drinking water. Despite all efforts, there are still almost 900 million people in the world who do not have access to clean drinking water. Not least for this reason, the development over the past decades has resulted in the fact that especially in hotter areas and in urban areas, more and more water is used from industrially filled bottles.

↑ Green landscape from the groundwater
29°56' N · 38°25' E (March 23, 1998 and February 25, 2003)

Wadi as-Sirhan is located in the north of Saudi Arabia near Jordan. Unlike in the 1980s, when only desert spread, large areas are now used for agriculture due to intensive irrigation. The patterns of circle irrigation are distinctive; these irrigated fields usually reflect a diameter of 3,280 feet (1 km).

35 km

17.5 km

0

Water for agriculture

For most people, personal use of drinking water is the primary use; however, with approximately 70 percent of the global water consumption used for farming purposes, agriculture clearly plays the main role in the overall picture. The largest portion by far accounts for the irrigation of agriculturally used cultivated areas.

In order to obtain a regular and high crop yield from growing cultivated plants, the cultivation area must be supplied with the appropriate quantity of water at any time. In many regions of Earth, especially in the dryer regions, precipitation amounts are too low or seasonally distributed too irregularly, that they would not allow for an efficient cultivation of the agricultural areas.

Therefore, all higher cultures developed systems very early on that allowed for a more regular irrigation of the cultivated areas in order to secure the food supply for a growing population. Today, the effects of agricultural irrigation are obvious even in satellite images. From extended canal and sluice systems that run through the fertile plains in East China, and the wet field cultivation of the rice terraces of Bali to the sophisticated systems or circle irrigation, there are many different forms of irrigation systems that become noticeable in the landscape picture in very different ways.

Due to the enormous water demand for agriculture and its vital importance for supplying people with food, the efficiency and sustainability of the use of water in this particular area has also gained more attention over the past few years. Irrigation forms, especially in dryer areas, share a common trait, in that a large portion of the brought-in water is lost due to evaporation. In addition to the loss of water, this results in dissolved minerals remaining in the ground, thus leading to salinization and decreased fertility. One of the important goals of irrigation is therefore to keep the evaporation rate low. This is accomplished by applying water as close as possible to the roots of the irrigated plants, as is the case in drip irrigation, rather than by large-area overhead irrigation. A decrease of the evaporation loss can, however, also be accomplished by cultivation in greenhouses.

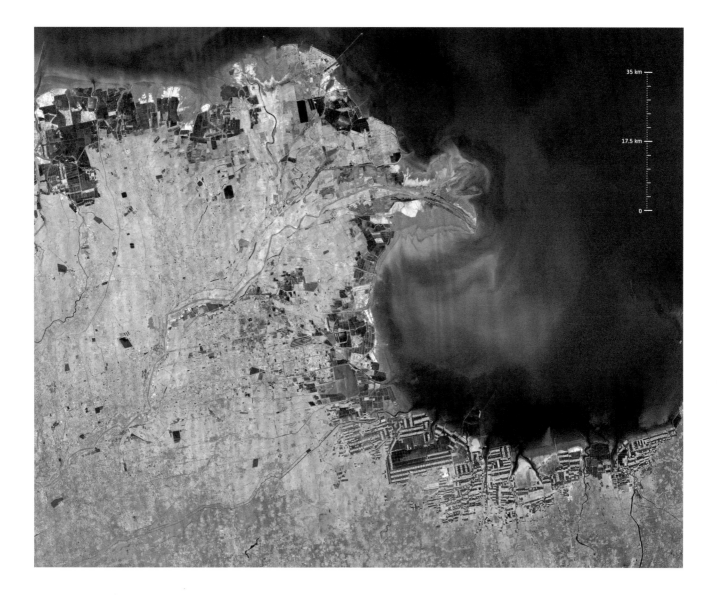

35 km

17.5 km

0

Like fish in water

In addition to collecting fruit and hunting for game, fishing is the oldest method of humans to obtain food. Until today, wild fish has remained one of the most important sources of protein for humans. In many coastal regions, fishery still forms the basis for food and accordingly also represents an important aspect in the cultural self-image of the population, resonated in many literature works, such as Herman Melville's *Moby Dick*.

The same as hunting and collecting, catching wild fish is only then sustainable if the system is not exploited. This means that every year, no more fish may be caught than is being produced by natural propagation. Due to modern fishery techniques, with their mile-long drift nets (which are illegal in many parts of the world) and deep reaching trawl nets, but also because of the large number of worldwide operating fishing fleets, the efficiency of fishing has long since reached a degree that is no longer sustainable. As a result, the populations of many species of food fish have been overfished in large areas of the oceans. This is already the case for approximately 90 percent of the population in the water bodies of the European Union. Some species of fish are even at risk of extinction. There is the

↑ **Farmed coasts**
37°33' N · 118°54' E (May 7, 1987 and June 4, 2009)

The mouth of the Huang He into the Chinese Gulf of Bohai is characterized by natural dynamics, as well as dynamics caused by humans. While the high sediment load of the river changes the shape of the mouth, aquaculture areas have spread in the adjoining areas along the riverbanks.

additional problem that about one-third of the catch is unwanted by-catch.

From wild fish catching to fish farming

Aquaculture provides a possible way out from the dilemma between the demand for fish and the limited resources in the water bodies of Earth. Similar to livestock farming, one changes from natural reproduction to breeding the required species, in which the operator must ensure the food supply and medical care of the breeding animals. In the small scale of long-known fish ponds, fish farming would nowhere near be able to measure up to today's requirements. Aquacultures have only been run for a few decades to an industrial extent, but they already produce more than one-third of fish marketed worldwide. During the past few years, considerable progress was additionally made with respect to the cultivation techniques in aquacultures, and as a result, it was possible to moderate the associated difficulties relating to food requirements, waste disposal and requirements for pharmaceuticals.

In addition to fish farming, aquaculture is also used for the production of shrimp, mussels and algae. Now, wide coastal areas of Eastern China are covered in aquaculture basins, which also characterize the satellite images of these regions.

Water and the industry

Approximately one-fifth of the total water demand is used by the industry. Here, water is used as a raw material, which is also found in the end product. Examples are numerous liquid products (such as washing detergent in which the "actual" product is dissolved in water), as well as many types of food, primarily, of course, all drinks.

Water is furthermore required in the industry for many physical and chemical processes. These include the cooling of machines and cleaning of facilities, but also diluting of salts and acids or the production of hydrogen and oxygen. Although the water consumption in the industry is relatively low when compared to agriculture, the consumption figures for individual products are still quite impressive. For example, you need approximately 105,670 gallons (400,000 l) of water for the production of one ton of paper, around 52,835 gallons

(200,000 l) for one ton of steel and 132,085 gallons (500,000 l) for one ton of plastic. Industrial facilities at least have the big advantage that the processing steps are executed in a concentrated space. It is therefore often possible to set up processing cycles in which the water is treated again and again and reused in order to thus minimize the consumption.

Virtual water and the water footprint

The difficulties in supplying all people with clean drinking water and the necessity for efficient and, above all, more sensible use of this resource, have lead to the development of the virtual water concept. Virtual water also includes water that is used for the production of all other utilized products and also their pre-stages, in addition to directly consumed water. It is intended to use this method to allocate the "imported" water consumption with the consumed products to the respective consumer and thus obtain a fairer representation of the resource consumption of people from different regions of Earth.

For example, this calculation provides for beef, a virtual consumption of 3,960 gallons (15,000 l) of water per 2 pounds (1 kg) of meat, the water required for the production of fodder representing the main portion. However, for more complicated products such as mobile phones or automobiles, the calculation of the virtual water consumption quickly becomes very complicated since the origin and water demand of all individual components must be tracked, which is hardly possible any longer due to the global interweaving of the economic processes.

In addition to the virtual water concept, a number of other terms were created to describe the efficiency of water utilization. For example, for agricultural use of water, a color system has been developed. "Green water" stands for the portion of rainwater that plants suck from the ground and evaporate, "blue water" has been taken from lakes and rivers or the groundwater for irrigation, and "gray water" is water that has been polluted during the production process, by pesticides, for example. Based on the ecological footprint and the CO_2 footprint, the generic term of "water footprint" was expressed around the turn of the millennium. It is based on the concept of virtual water and specifies the consumption of virtual water per capita or per country.

Landscape in a grid view →
37°35' N · 100°40' W (July 19, 2013)

West of Dodge City, in the southwest of Kansas, you will find a wide landscape that is characterized by the grid pattern of the land reclamation that was carried out in the 19th century. Basis of the grid is a section that comprises 1 square mile (approximately 2.6 sq. km). Besides the naturally irrigated square fields, you can identify the round and clearly more densely vegetated fields of the central pivot irrigation systems.

0   4 km   8 km

Fertile high valleys
36°33' N · 103°46' E (June 17, 2010)

Xichazhen is located in the foothills of the Loess Plateau in the Chinese Gansu province. Due to the dry climate in almost 6,560 feet (2,000 m) above sea level, the agricultural use is almost exclusively limited to the wet valley plains, while the rugged, unstable slopes are not well suited for this purpose.

0   250 m   500 m

Rice terraces on the hills
16°51' N · 121°06' E (September 22, 2013)

In the mountainous interior of Luzon, the main island in the Philippines, near Hingyon, widespread terraces for wet field cultivation of rice were created at almost 3,280 feet (1,000 m) above sea level. On the flat ridges of the hills, the fields are allowed to be relatively large as compared to the steeper terrain.

Salmon in cool waters

41°49' S · 73°06' W (January 25, 2011)

The island Chidguapi is located in the Golfo de Ancud near the South Chilean city of Calbucu. Since the cold water with high oxygen content is suited really well for salmon farming, the entire island, as well as other coastal areas of the region, is lined with aquaculture facilities.

Fish ponds in the Mekong Delta

10°21' N · 106°42' E (February 17, 2010)

Gò Công is located near the Vietnamese capital Ho Chi Minh City in the delta of the Mekong. This area is characterized by a high population density, which is based on the high degree of fertility of the ground. Numerous farms, which use the abundance of water not only for irrigation, but also for fish ponds that intersperse the fields and orchards, are located along the dense network of canals and roads.

FISH IN THE GARDEN

People have been using fish ponds for many years in order to ensure their supply of fresh fish for their nourishment. In China, these artificial water bodies existed as early as 4,500 years ago, and in Europe they were already used by the Romans and later especially for monasteries and properties of monarchs. Fish species like carp or pike with a low oxygen demand and that live by nature in standing bodies of water are especially well suited for farming in basins and ponds. One fish that has conquered the European market is the shark catfish, which originates from its natural habitat in Southeast Asia and can reach a length of 3 feet (1 m) in the wild. Fishing significantly reduced stock of this fish species over the last few decades so that the shark catfish was placed on the Red List of endangered species. Farming of this undemanding fish is therefore now customary, especially along the lower stretch of the Mekong River, where it is kept in flooded rice fields and numerous fish ponds specifically created for this purpose.

Nourishing basin

44°41' N · 1°03' W (February 16, 2009)

The Bassin d'Arcachon is located near the western French city of Bordeaux. The flat bay is strongly exposed to ebb and flow, and is also known for Europe's largest sand dune — Dune du Pilat, located in the south of its entrance. As early as in the first half of the 18th century, basins for salt production were built near Audenge and were converted to fish farming basins in 1843.

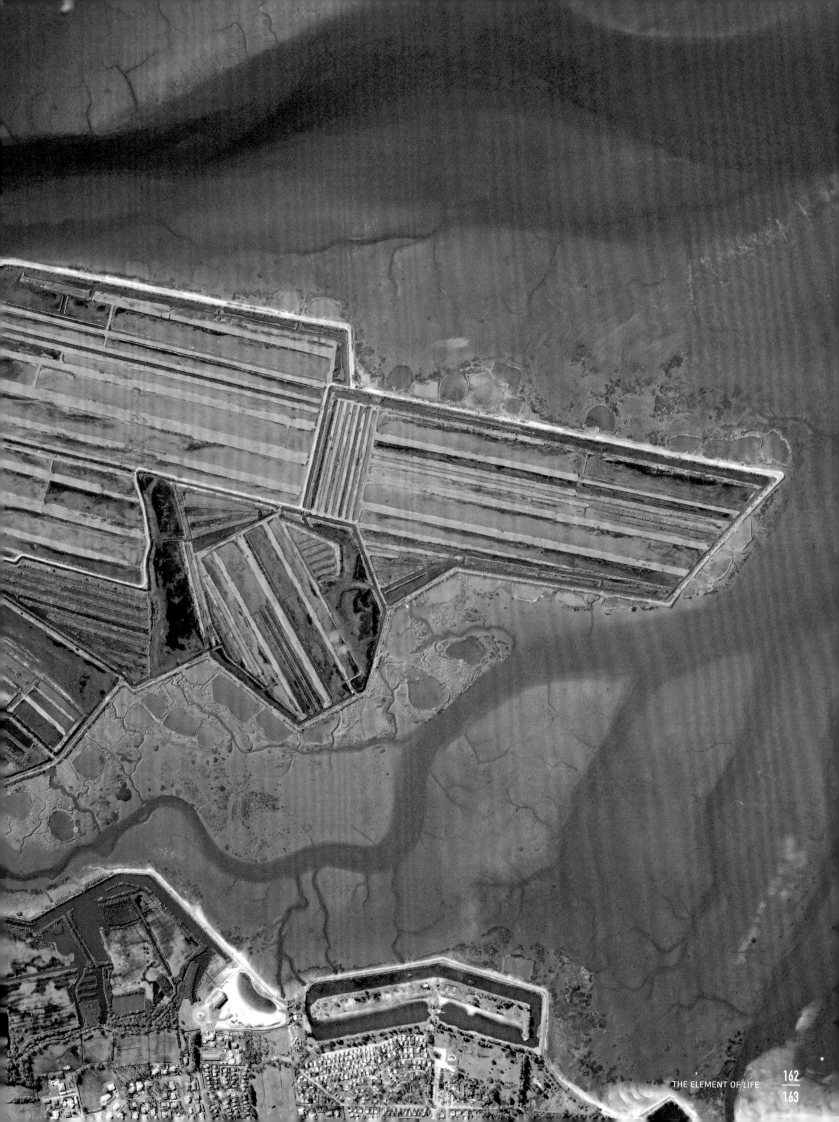

As on many coasts of China, aquaculture areas
represent an important economic factor also at
the Gulf of Boha. They are a major contributor to
feeding the population and serve, to some extent,
as a pre-stage for land reclamation in areas close
to the coast. Due to its close vicinity to the capital
city of Beijing, the Gulf of Boha is one of the world's
seaways with the highest volume of traffic.

WAYS OUT OF COMMERCIAL FISHING

Aquaculture is one answer to the increasing overfishing
of large areas of the oceans, which affects almost all
exploited species of fish. Although fish have been farmed
since ancient times, for a long time, it was far more economical to catch wild fish. The development of efficient
catching methods also contributed to this, which led to
the situation described at the beginning. The number of
aquacultures has therefore increased rapidly over the past
few decades. In accordance with the Food and Agriculture
Organization (FAO) of the United Nations, approximately
three million tons of fish were produced in aquacultures in
1970; in 2012, it was already more than 66 million tons,
which equals a production value of $138 billion. At least
23 million people were employed in this sector, 90 percent of the production being carried out in the Asia-Pacific
region. The same as all other forms of intensive cultivation, aquacultures also have their downside. This includes
mainly the destruction of natural habitats, the intensive
use of pharmaceuticals and problems with waste.

AT THE TURNING POINT
Overfishing endangers the future of Earth

Without the biological diversity in the oceans, human life on Earth would be impossible. But our reckless use of the oceans is threatening to destroy this foundation. A quick change of course is urgently required.

Our home planet, Earth, is an ocean planet. The oceans cover over two-thirds of the Earth's surface and with a volume of over 320 million cubic miles (1.3 billion cu. km) they bind approximately 97 percent of all of the Earth's water reserves to a world without barriers.

Life exists only where there is water: And so it really comes as no surprise that animals, microbes and plants occur nowhere else in such large numbers and such bizarre forms as below sea level — the largest inhabitable space in the biosphere by far.

From the 26 known anatomic blueprints of biology, 25 can be found in the sea, but only 12 on the continents. The largest giants of the living world, whales and siphonophores swim far away from the coasts, the same as myriads of tiny plankton animals and algae, some of which are only 0.00001 inches (0.0002 mm) in size, in addition to glittering mythical creatures, winged octopuses, medusas and frogfish, butterfly blennies and dragonlike creatures.

One single gallon of sea water can be home to thousands upon thousands of organisms. We humans are beginning to discover the diversity of life in the undersea cosmos. It is now believed that more than 10 million, maybe even 100 million, species of fish, cepha-lopods, snails and marine bacteria swim, crawl and drift though the depths still undetected.

About the incredible value of our oceans and their threat

We only know a tiny fraction of the oceans. Despite all progress in developing diving equipment, the oceans are still too vast, difficult to explore, dark, cold and filled with water with hostile pressure. Every expedition into it makes it even clearer: We humans are on the margins of this water world. But without the diversity of life in the ocean realm, our own species could not exist either.

We need the oceans — as pollutant filters and climate buffers, as sources for food, medication and biotechnological, as well as for cultural inspiration. The economic value of coral reefs alone is estimated by experts to approximately $327 billion per year, and that of the oceans in total to even at least three sextillion.

We breathe the air of the oceans, because as photosynthesis powerhouses, the tiny algae that float in the upper levels of the water produce more oxygen than all terrestrial plants together. Corral reefs and forests of seaweed protect the coasts. For almost half of the human population, fish and seafood are the most important protein basis of their daily food.

Approximately 100 million tons of crabs, fish and cephalopods are caught in the oceans every year. With the growing world population, greed continuously increases. However, the oceans will not be able to satisfy this demand for much longer. According to estimates from the FAO, today almost one-quarter of the world-wide fish stocks are considered "overexploited."

They are hunted down so relentlessly that they cannot recover. Another 30 percent are at the verge of overfishing. In some regions, like the Mediterranean Sea for example, scientists estimate the percentage of overfished stocks to have reached 80 percent. The big nomads of the open oceans are especially affected by overfishing: Tuna and swordfish, halibut and sharks for example. In only five decades, most species of the large carnivore fish have been decimated by 90 percent.

The profits from this hunt of the hunters of the blue depths are distributed extremely unequally. Only one percent of the worldwide fishery fleet brings to land around half of what they caught at sea. Captains of super trawlers hunt by means of satellites and radio buoys for the hiding-places of the shoals of fish. They cast fishing lines that can be up to 60 miles (100 km) long, along with nets into which several jumbo jets would fit next to each other.

And though catching techniques are increasingly improving, the hooks and nets remain empty more and more frequently for some years. The oceans are at the limit of their creative power. If we do not curb overfishing right away, fishermen all over the world are endangering their own future. Even more: The biological network of species in the oceans as a total is on the verge of tipping over — with unforeseeable consequences for our own existence, but mainly for the existence of our children and grandchildren.

Time to rethink

For far too long, we humans have considered the oceans as inexhaustible. Now, however, not alone among environmentalists, but also in the economic sector and politics, the conclusion establishes itself more and more that we must radically change our previous use of the resources of the oceans. For example, New Zealand, Norway and Canada have been regulating fishing with strict laws for several years. Meanwhile, the EU is also striving for a more sustainable fisheries policy. The fishing quota is to be more restricted and exceeding the quota shall be punished more severely. And it will no longer be allowed to throw "by-catches" (all those ocean creatures that have been killed unintentionally by nets and hooks) back into the sea. They must be brought back to the harbor. The economic loss that captains incur due to useless filled loading spaces is intended to entice the fishermen to catch their goods more considerately and more selectively in the first place.

However, more than these regulations, scientists consider another instrument vital to secure shoals in the oceans in the future: They urgently request that more protected areas in the oceans are established so that exploited stocks can recover. A network of such ocean parks that interlink particularly valuable "focal points" for the biological diversity could considerably moderate the damages from over-fishing within just a few years.

Not even 3 percent of the worldwide areas of oceans are currently under protection; this is six times less than on land. In accordance with many experts, this figure should be at 20 percent at least. Some countries like Germany, Australia or the United States have already fulfilled this goal in their bodies of water close to the shores; however, the actual protection status of the declared reserves differs substantially.

Nets, harpoons and hooks are not actually forbidden by any means in all ocean parks in the world. And yet, some examples show again and again how fast the marine creatures can regenerate in strictly regulated protection areas. For example, in New Zealand, the waters around the island group Poor Knight Islands were reserved for nature in 1998. Trawler captains and sport fishermen are no longer allowed to catch anything here; instead, divers from all over the world marvel at the sea chubs and anthias that pass through the canyons of the underwater world that are richly covered in corals. The stocks of some species of fish have already multiplied eight-fold since the establishment of the protection zone. Numerous sea creatures that had disappeared from here have returned. And the people at the coasts also profit from such refugia for fish.

The villagers near the Spanish Medes Islands, for example, have an income of approximately $6.8 million per year from diving tourism, since these waters of the Mediterranean Sea have been declared ocean protection zones — this is 20 times as much as through fishery before.

Ocean protection zones prove this: Enormous self-healing powers are slumbering in the oceans. We only need to give them a chance to develop. We owe it to the next generations.

Lars Abromeit, born in 1974, is an editor and reporter of the GEO magazine in Hamburg. He accompanies scientific expeditions to deserts, rain forests, mountains — and often to the sea. Lars Abromeit has studied law and biology and has worked as a freelance author among other countries in Australia and the Mascarene Islands. He has received many awards for his reportage work, including the Henri Nannen Prize. Together with the photographer Solvin Zankl, he has written the book Ozeane — Expedition in unerforschte Tiefen (Frederking & Thaler) about the impressive creatures in the sea.

Note from translator: translates to:

Oceans – Expedition into unexplored depths

"The biological network of species in the oceans is on the verge of tipping over — with unforeseeable consequences for our children and grandchildren."

Protein source — the ocean

23°45' N · 117°34' E (August 8, 2011)

The coastal area between the flourishing Chinese
cities Xiamen and Shantou provides a flat transition
to the open sea of the Taiwan Strait — ideal prerequisites for an intensive utilization by aquacultures that
are operated correspondingly on a large scale. Beside
fish, mainly seaweed is being farmed.

0   150 m   300 m

An abundance of fish
13°14' N · 103°49' E (April 23, 2004)

As the largest freshwater lake in Southeast Asia, the Tonlé Sap in Cambodia plays an important ecological and economic role and, is home to the Mekong giant catfish, the largest living freshwater fish, as well as many other species of fish. Fishermen who live in the villages (some of which are floating) around the lake, use the enormous abundance of fish in the lake.

0   150 m   300 m

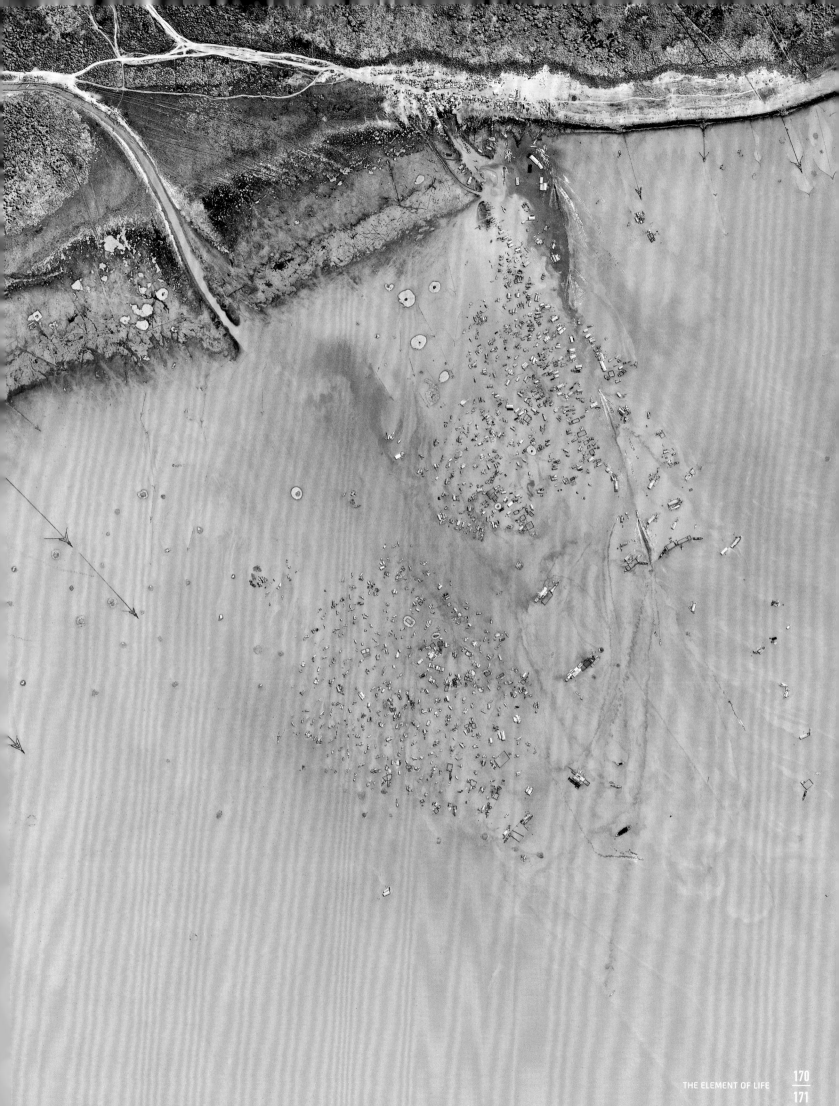

Refreshment from the mountains
31°42' N · 4°11' W (December 27, 2010)

The Moroccan village Aoufous is located at the southern flank of the High Atlas in the valley of the Ziz River.
Surrounded by the dryness of the extensions of the
Sahara, the activities of humans are concentrated at
locations with access to water that mainly originates
from the mountains. Suitable areas along the river
are used for agriculture, while settlements are located
behind these areas because of the flood protection.

0   300 m   600 m

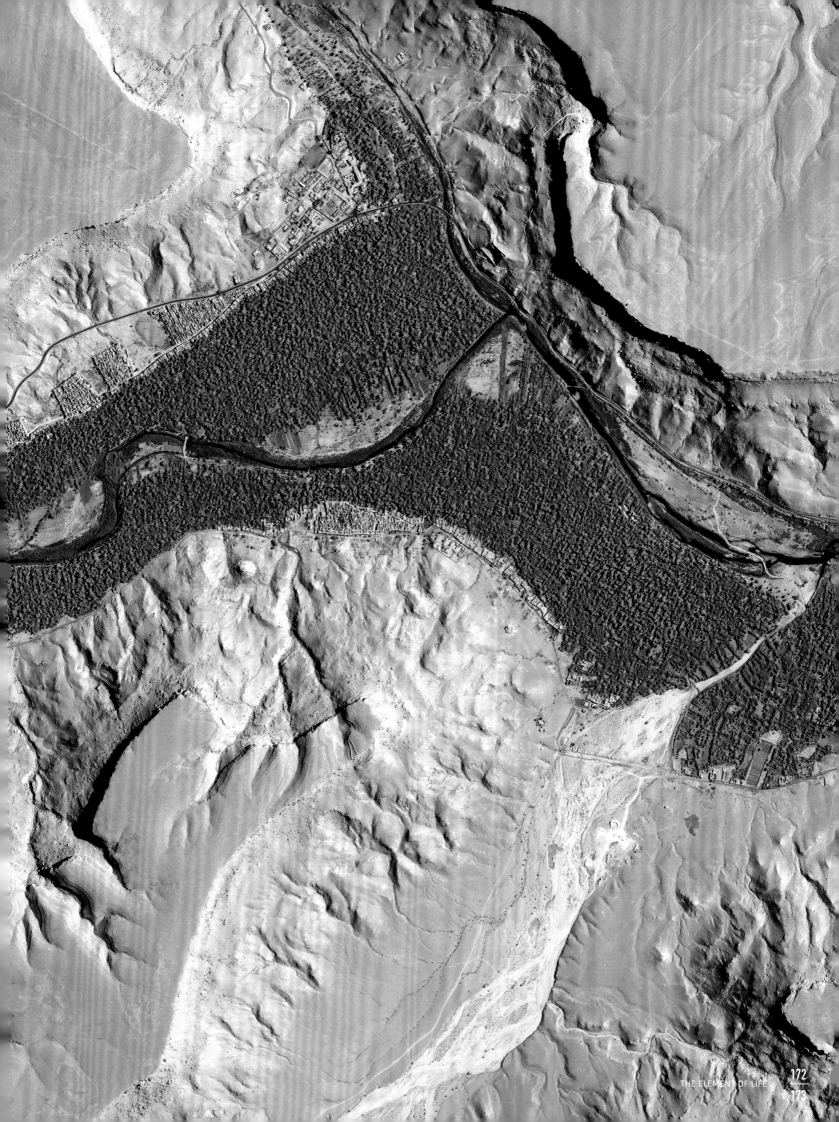

Fatally miscalculated
31°26' N · 31°51' E (October 11, 2013)

As a result of drainages, a large part of the brackish lake Bahra al-Manzala in the northeastern area of the Nile Delta was drained and thus lost for fishery. However, the crop yield on the newly obtained areas fell short of expectations because of the salty soil. Salt is also produced in close vicinity to the city of Damietta.

WATER: A CAUSE FOR CONFLICT

Water is a resource that can be used in extremely versatile ways: from fish farming to irrigation, from energy production to transportation means, from the use in production processes to the use as cleaning agent and as food, to name just a few examples of the wide range. Often, conflicts arise between the different users, for example, when bodies of water that are used as a source of drinking water are polluted by industrial plants or when water is used for irrigation, which is then missing downstream for energy production. Particularly problematic in this connection are mining operations, pollution from which can make water unusable even far away from the party responsible.

VULNERABLE BLUE
Conflicts between humans and the water

Tons of plastic waste that is hardly degradable, the chase for valuable resources and climate change are afflicting our oceans. Is it too late for help? No — but we must finally act.

The oceans are boundless, fascinating and threatening at the same time. Anyone who has floated through the exotic underwater world, past colorful coral gardens, glittering shoals of fish and thick algae forests, will never lose the enthusiasm for the oceans of our world. Influencing cultures were developed at their shores and scientists searched for new knowledge.

And still we are just scratching the surface of the marine ecological cosmos. But that does not stop us from pushing the industrialization of the oceans, exploiting their resources and using them as garbage dumps. We run the risk of losing entire habitats and thousands of species before we even discover them. We are cutting the life-giving water from the Blue Planet.

Oceans have always provided us with food, as well as traffic and trade routes. Almost 70 percent of Earth is covered in water. The productive shelf seas, undersea mountain ranges and the blue to deep black open water form more than 90 percent of the animated biosphere of our planet. Over 250,000 different species have been described today and new ones are discovered almost every day.

The three-dimensional world full of fascinating creatures starts just several feet below the water surface. Fable-like creatures, like blind lobsters and vampire squids, live in the eternal darkness of the deep sea. It is a phenomenal universe that must be protected.

Water as a garbage dump

The list of threats is long, but none has become as obvious as the threat from plastic garbage. It only took 100 years to change our plant irrecoverably, because plastic does not decay; it only dissolves into increasingly smaller particles — one plastic bag over 25 years and one plastic bottle over 450 years. The ecological consequences of this durability are dramatic: Dolphins and fish get caught in old fishing nets, sea turtles mistake plastic bags for jellyfish and albatrosses mistake lighters for fish. Once the garbage has been swallowed, the animals die of starvation with a full stomach or die from internal injuries.

Scientists are getting increasingly worried about the smallest representatives of the deadly plastic armada. Microplastics, released from the breakdown of larger garbage pieces or from direct discharge from cosmetic products. Today, you find plastic in mussels, crabs, fish and gray seals. Cause for toxic ingredients are also cause for great concern (like the hormone-changing bisphenol A, softeners or flame retardants). And only a few years ago, scientists discovered that cancer-causing insecticides like DDT attach to the plastic particles and accumulate via the marine food chain. The implications for humans are also unknown.

Neglected for a long time, society has finally recognized the problem with plastic waste. It quickly became evident that the oceans can no longer be cleaned completely and that waste collection initiatives, like Fishing for Litter, can only limit the damages locally. The estimated quantity of 140 million tons of plastic waste in the oceans is too large. It is necessary to stop the influx of up to 10 million tons per year. The solutions can be found on land, in the waste and recycling industry, in an improved resource efficiency and innovative product design. The time for action is running out.

Chasing for resources

Besides the environmental pollution, there are also the various utilization interests that afflict the oceans. Sand and gravel, oil and gas have been awakening economic desire for a long time, as well as the mining of diamonds, titanium, silver and gold in the flat shelf areas. The oceans are full of treasures. Until a short while ago, large-area mining in areas far from the coast was not economical. As a result of technological progress and increasing commodity prices, interest in mining in these areas is rising; the ocean mining business seems more lucrative now.

In the center of the resource rush are the manganese nodules (mostly at water depth below 13,125 feet/4,000 m), but also gas hydrate and cobalt crusts. Germany does not want to be left behind in this development and has already secured mining permissions for manganese nodules for approximately 29,000 square miles (75,000 sq. km) of the Pacific Ocean bed from the International Seabed Authority. The treasured nodules contain the metals manganese and iron, but also copper, nickel and cobalt. These are metals that are needed in the steel production and electrical industry.

However, mining resources in the deep sea is associated with high environmental risks. The extraction causes noise and vibrations. This might disturb dolphins and whales. The German mining area is located approximately 620 miles (1,000 km) off the Mexican coast, south of the Baja California, the nursery ground for the Pacific Gray Whale. In addition, all animals in the ploughed-through area that are not fast enough to escape will die (worms, snails, sea cucumbers, etc.). And the industrial exploitation of the resources can release toxic heavy metals. Mining in the long-lived and sensible ecosystems, such as the Arctic and the Antarctic are particularly controversial. In this context, it becomes forgotten that the oceans are already threatened by overfishing, pollution and the effects of climate change. Environmental and conservation organizations are campaigning for a moratorium for ocean mining, a UN regulation for the sustainable use of the high sea and more protected areas.

Climate change as a risk factor

Besides the dispute about the protection and use of the marine biological diversity, another manmade problem is threatening to fundamentally change the oceans: climate change. Warmer and more acidic are the alarming prognoses. The oceans are the most important stabilizing power of our climate system; the huge bodies of water store and transport heat, and influence our weather and the climate through their global circulation systems. They absorb 50 times more greenhouse gases than the atmosphere and approximately 30 percent of the carbon dioxide that is produced by humans annually.

The thermal expansion of the water and melting of the ice caps considerably accelerate the rise of the sea level. Floods, large catastrophic tidal floods and the loss of habitat for humans and nature are threatening as apocalyptic consequences. This is accompanied by an increasing acidification of the oceans. During the past centuries the pH level has decreased by 1.1 units on average. For many occupants of the oceans, this is like a death sentence; because acids dissolve lime, organisms that form lime, like corals, mussels and many algae, are having it increasingly difficult.

The effect of shipping on climate change is often underestimated. Over 90 percent of the international transport of goods is processed by sea. Due to their high freight volume, transports via container ships are considered extremely efficient — however, they are not environmentally friendly.

In addition to tens of thousands of freight ships, hundreds of cruise ships travel the oceans. Off populated coastal areas, they also mutate to real polluters. They also use heavy fuel oil, a toxic leftover from petroleum refining, as fuel. It can change the genetic make-up and cause cancer. Toxic substances are released during its burning, including soot, sulfur oxides and nitrogen oxides. In accordance with the World Health Organization (WHO), these toxic ship emissions are responsible for up to 50,000 premature deaths in Europe. Soot is increasingly having an effect on the global warming of Earth. The black particles blow all the way to the Arctic where they accumulate on the ice. And yet, the solutions have been on the table for a long time: efficient catalyzers, soot filters and innovative ship engines. But these solutions cost money. And neither the shipowners nor the cruise passengers are willing to pay this price.

All these threats must not result in us resigning. Fighting for a better protection of the oceans, for the future of porpoise, tuna and albatross is worth it. Earth will always remain a Blue Planet. And we are all responsible for the fascinating life in the endless, vulnerable blue.

Dr. Kim Cornelius Detloff is a marine biologist and worked for several years at the Institute for Marine Biology on the Italian Giglio Island after his studies, before he became engaged as a campaigner at the International Fund for Animal Welfare (IFAW) and as a political-scientific consultant at the Bonn Convention (CMS). Today, Detloff is head of the Marine Protection department at the German federal association NABU (Nature and Biodiversity Conservation Union).

"The list of threats is long. But none has become as obvious as the threat from plastic garbage."

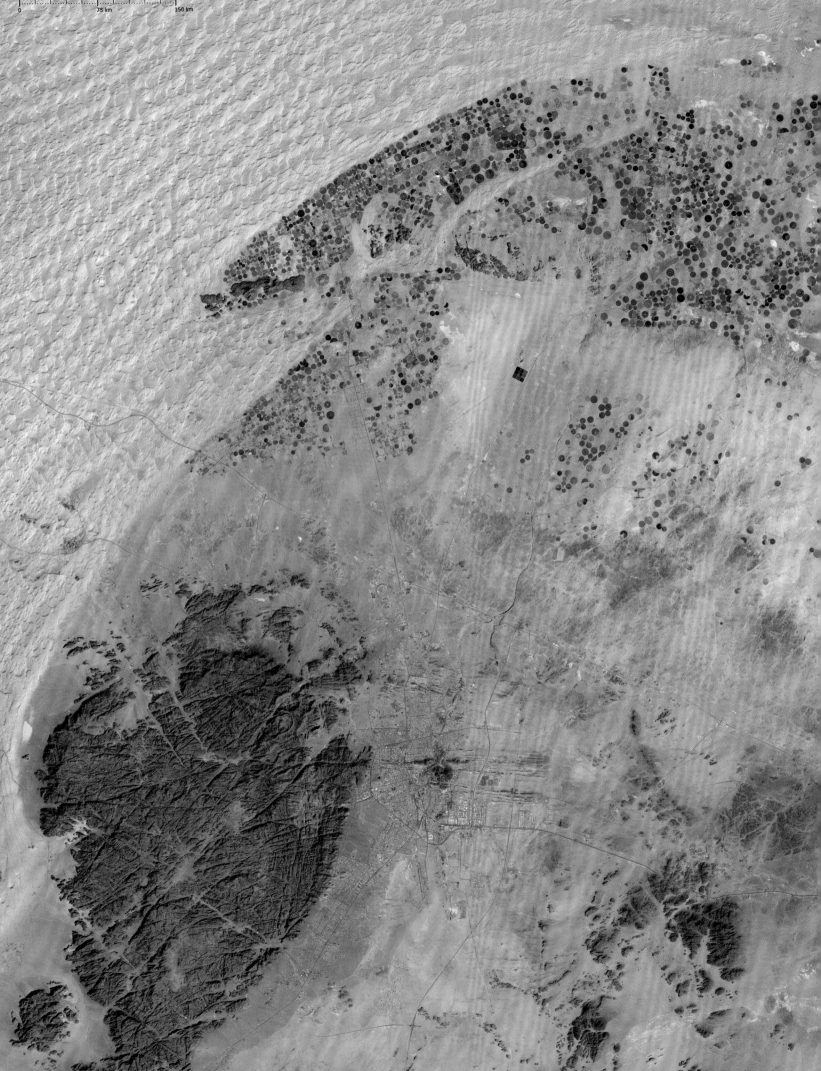
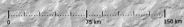

Green fields in the desert
27°39' N · 42°10' E (February 10, 2014)

North of the city of Hail provides a good example to illustrate the development of Saudi Arabia from an import country to an export country for agricultural products. By exploiting the fossil ground water reserves, the cultivation areas in the desert were drastically expanded. Salinization of the soil as a result of intensive irrigation is intended to be minimized by selecting appropriate crops and irrigation methods.

Useless water
33°01' N · 115°35' W (May 15, 2013)

The Imperial Valley, which is intensively used for agriculture (for livestock breeding and fattening, and for the cultivation of fruit, vegetables and cotton), is located between the Salton Sea and Mexicali-Calexico. Water from the All-American Canal is brought in from the Colorado River — the water from the closed Salton Sea would be too salty.

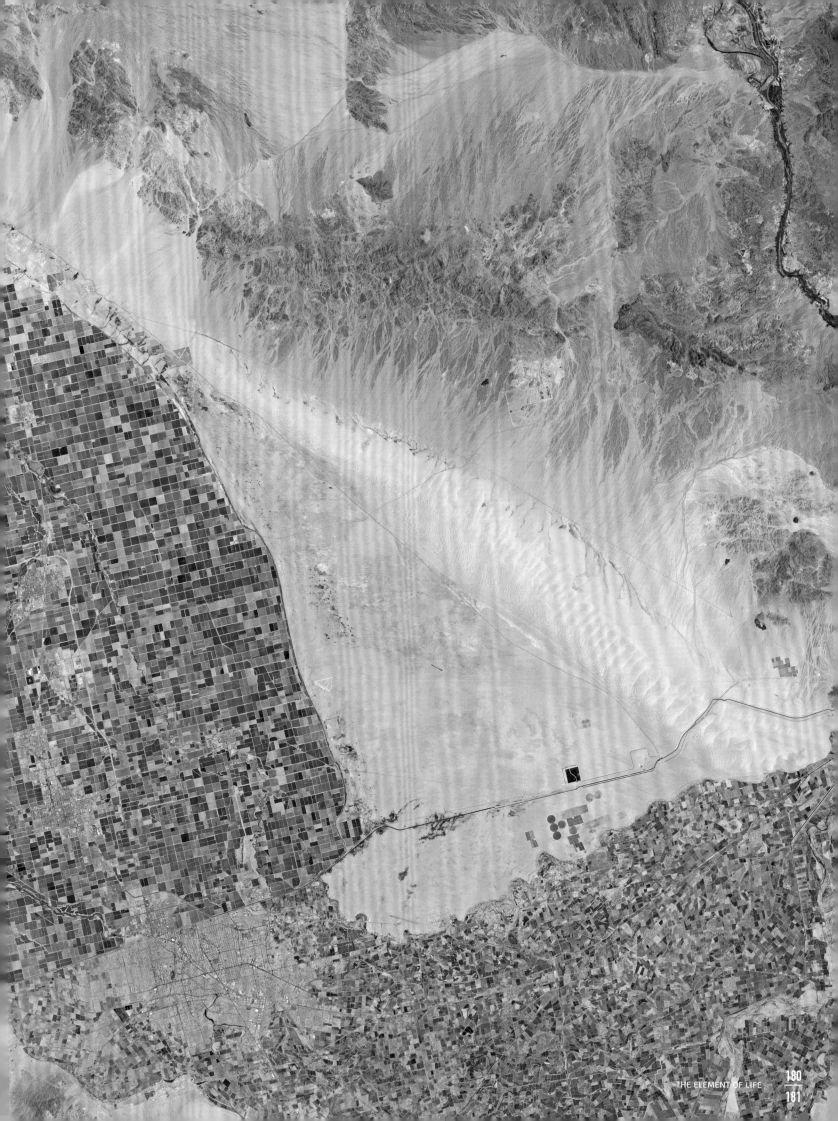

Getting more and more thirsty
29°54' N · 56°51' E (July 07, 2010)

The dry climate around the city of Bardsir in the Iranian province Kerman requires the irrigation of cultivated areas. Traditionally irrigation was accomplished by means of canal systems of the underground qanats, which transport water from the mountains to the plains. However, deeper lying groundwater levels are also used as a last resort in order to satisfy the increased demand for water. Both systems are shown in the satellite image: Qanats with their mile-long chains of shafts in the ground, as well as the expanded fields with modern circle irrigation.

DISADVANTAGES OF IRRIGATION

Irrigation of cultivated areas is one of the prerequisites for reliable yields and subsequently, for efficient agriculture. Especially in dry climate zones, agriculture without artificial irrigation would be unthinkable. Besides thousands-of-years-old methods to use the available water in a more or less sustainable way, procedures that appear more efficient at first glance have established themselves during the course of the industrialization of agriculture. As a result, huge cultivated areas are now artificially irrigated. In addition to the exorbitant water consumption, the salinization of the soil poses the main challenge. The minerals that are dissolved in the water remain behind during the evaporation process and accumulate in the soil. However, the increasing salt content lowers the fertility of the soil so that agricultural use of the areas sooner or later becomes impossible. Refined irrigation procedures enable a minimization of these damaging side effects; however, they require awareness of their importance.

Growth under foils
28°09' N · 15°39' W (September 13, 2012)

Although the climate of the Canary Islands is very pleasant, the water on Gran Canaria is distributed very unevenly and must be used efficiently. In order to also minimize the impact of the often severe winds on the yield, fruit and vegetables, including bananas and tomatoes, are mostly grown in foil greenhouses.

A CONTESTED COMMODITY
Drinking water shortage calls for intelligent solutions

Access to clean water is becoming a problem for increasingly larger portions of the growing world population. A pilot project in the north of Namibia shows how one can use the precious water in a sustainable way.

Today, the world population is around seven billion people and in 35 years, we will presumably reach nine billion. Demands, family structures and lifestyles are changing globally. People are becoming more and more mobile. They are increasingly flocking to cities and metropolises. These changes impact our use of water. In many regions of the world, the average annual water withdrawal already today exceeds the renewable quantity of available water.

Due to the population dynamics, climate change and globalization, an intensification of water shortages must be expected. Accordingly, an increase of conflicts around water that is becoming scarcer seems to be inevitable. In the peace and conflict research of the environmental sciences, these developments have triggered a dialogue as to whether these global trends also result in an increase of the potential for violent disputes around the resource of water. For example, cross-border disputes around water resources are known for the neighboring countries of the Nile River or for the Jordan River Basin in the Middle East. In these regions, ground and surface waters cross international borders, control of which is often hard-fought

over. However, cross-national political policies allowed for an improvement of the cooperation between the conflicting parties. So far, feared "water wars" failed to materialize — on the contrary, in these cases, the scarce water resources were the cause for cooperative solutions between the countries. However there are also cases — especially at a national and local level — where conflicts around scarce water resources turned into violent conflicts. Such developments are especially probable when rules, institutions or agreements are scarce and the power of controlling the water resources serves power political purposes.

"Blue" and "green" water

Globally, the use of freshwater resources has increased eightfold in the last century, and increases by approximately another 10 percent every decade.

Water plays an important part in the production of food, and accordingly conflicts around scarce water resources often not only represent a threat to the drinking water supply, but also to the basic supply of the population with food from their own country. This applies in particular for arid and semi-arid regions in developing countries.

Science distinguishes between different types of water: "Blue" water refers to the water in streams, rivers, lakes and underground groundwater. It can be used for irrigation agriculture, for the industry and for settlements and is used worldwide — with large regional differences — by approximately 70 percent by agriculture, approximately 20 per-

cent by the industry and 10 percent by private households. "Green" water refers to the water that falls in the form of rain, infiltrates the ground, is absorbed by the plants and is stored in the soil moisture.

In arid regions with little rainfall and where the evaporation in total is higher than the precipitation, discussions are being conducted as to whether the technical provision of the blue water by means of dammed lakes, rerouting of rivers and exploitation of groundwater is sufficient in view of the global climate warming and the growing demand for water. The question arises whether these measures that are targeted toward an expansion of the water supply are maybe rather part of the problem that should be solved and part of the problem that may exacerbate conflicts. That is because large canals that divert the water, or artificial irrigation with groundwater, often produce large losses of precious water and large-area salinization of the soil.

Intelligent resource management

This can subsequently lead to new distribution conflicts and utilization competition around scarce water resources — between rural and urban residents, between different economic sectors and private households. In order to moderate this and ensure the supply, water resources must be used more intelligently and more efficiently. In many countries, there are now innovative approaches for a sustainable use of water resources.

One example is the Cuvelai Etosha Basin in the north of Namibia. This area is one of the driest regions in southern Africa and almost half of the country's population has settled here. The impact of the climate change, population growth and rural exodus increase the pressure on the water resources that are scarce anyway. Weather extremes such as distinct dry spells, but also severe rainfalls and floods, determine the lives of the people.

Here, the research project CuveWaters has the aim of achieving a lasting improvement of the livelihood by means of regionally adjusted integrated water resource management. The intention is to provide the population with secure access to clean water in the long term, and that poverty and diseases are reduced in the future.

Solutions are developed that flexibly adjust to the demand and that use different sources, types and qualities of the water for different purposes. Pilot plants that were adjusted to the local conditions were built in different municipalities. Solar-powered desalination plants treat the groundwater and deliver fresh drinking water. Modern storage systems are used to collect the rain and flood water during the summer months. They facilitate the irrigation of the fruit and vegetable gardens all year round. This allows several families to cultivate vegetables, sell them at local markets and subsequently tap new sources of income.

In another pilot plant, an innovative energy-efficient sanitary and waste water concept is being implemented. Waste water from the settlement is conducted through canals to a treatment plant. The treated water is used for irrigation together with nutrients. With it, a co-op of farmers cultivates agricultural areas and sells their products at local markets. At the same time, biogas is produced in the course of the waste water treatment, which is used for power and heat generation. This solution offers new per-

spectives for rural settlements, but also for fast growing parts of cities.

In the end, cooperation is decisive for success

One important prerequisite for the success of such initiatives is the close cooperation between the different partners and the inclusion of the local population. Together with political decision makers, partners in practice and residents, the plants are developed and tested, adjusted and operated in CuveWaters. At the same time, the local users acquire the required technical skills for the operation of the plants. Therefore, not only is education improved and unemployment reduced, but the self-responsibility for the local people and "their" water supply systems is strengthened. An improvement of the drinking water supply is connected with a higher food security, an improvement of health and a more effective fight against poverty. In view of the globally growing pressure on the resource of water and the subsequent increasing risk of conflicts, the results of such research are of great importance. They provide valuable information regarding possibilities of a solitary use of scarce water resources and a sustainable water supply also in other regions of the world that deal with similar problems.

Dr. Diana Hummel is a political scientist. She is member of the institute management at the Frankfurt ISOE Institute for Social Ecological Research and conducts research concerning the topic of population development and supply systems.

"Climate change, population growth and rural exodus increase the pressure on the scarce water resources."

Lake Yoa, the second largest of the group of the Lakes of Ounianga, is located approximately 125 miles (200 km) southeast of the Tibesti Mountains in the middle of the Sahara. Apart from the fact that these lakes are an important water source for caravans, they are also interesting if you consider them a climate archive, since the development of the climate in the previously far greener Sahara can be tracked from its sediments.

0   200 m   400 m

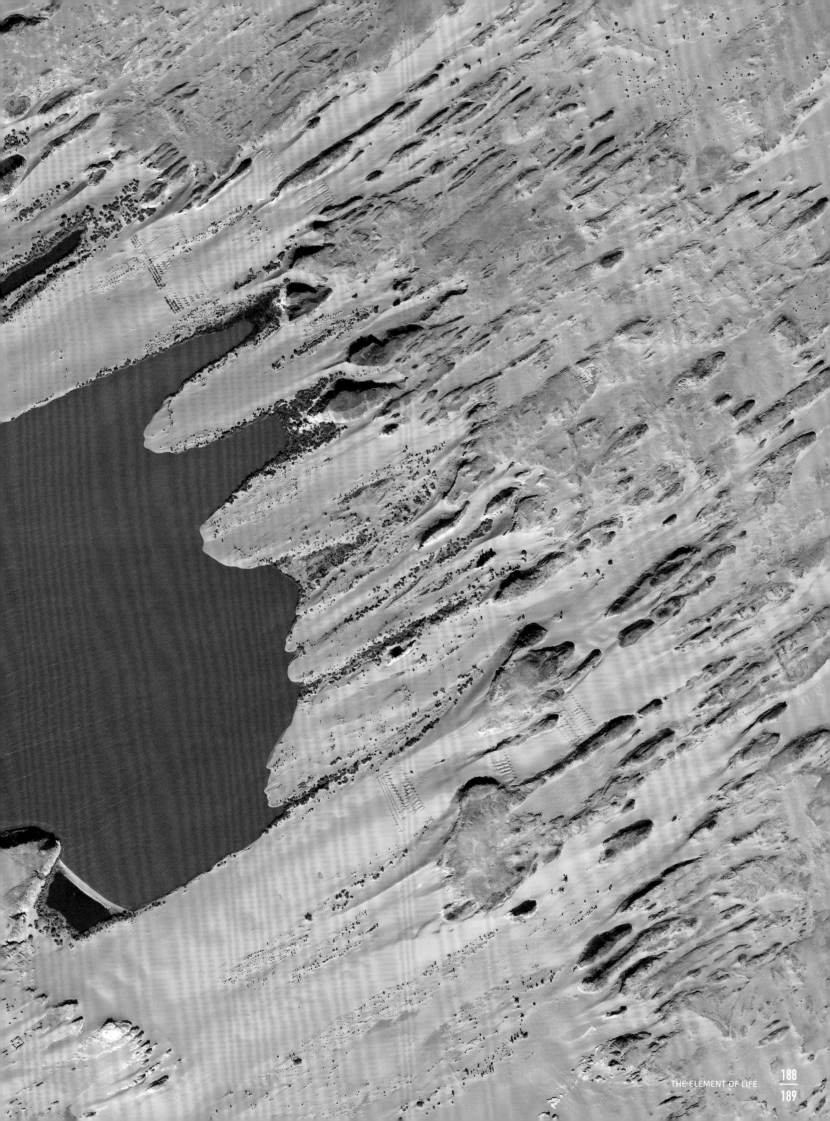

Floating gardens

20°33' N · 96°53' E (October 11, 2013)

With an area of over 39 square miles (100 sq. km),
Lake Inle is the second largest lake of Myanmar. Some
species of fish are only found in this lake. Due to the
extreme seasonal differences of the lake's water level,
many buildings were constructed on stilts along its
edge. The residents of these buildings are fishermen
and cultivate vegetables in floating gardens.

300 m

150 m

0

Shrimp against mangroves

12°55' N · 87°15' W (November 7, 2010)

Similar to many other coastlines of the world, large areas of valuable mangrove forests were destroyed at the Gulf of Fonseca in favor of shrimp farms. The mangroves that are prevalent on flat tropical coasts grow roots in the ground of the shores and protect them against erosion, but they are also important for the propagation of numerous species of fish and as a habitat for birds and reptiles.

Land from the sea

53°20' N · 5°46' E (April 20, 2011)

Hallum and Marrum are villages in the Dutch province of Friesland that have been built on old land reclamation areas. Located not far from the coast of the tideland, they were built around slightly elevated village centers. New, usable land areas that are interspersed with ponds arose as a result of dike construction and drainage of the tideland.

Agriculture under water
34°13' N · 126°33' E (December 3, 2003)

The flat bodies of coastal water off the southern tip of South Korea provide ideal prerequisites for operating aquacultures. Wide areas of the ocean bed between the islands around Sinan are covered in underwater beds in which seaweed is being farmed. During low tide, some of the facilities protrude from the water and are then accessible for maintenance work.

0   200 m   400 m

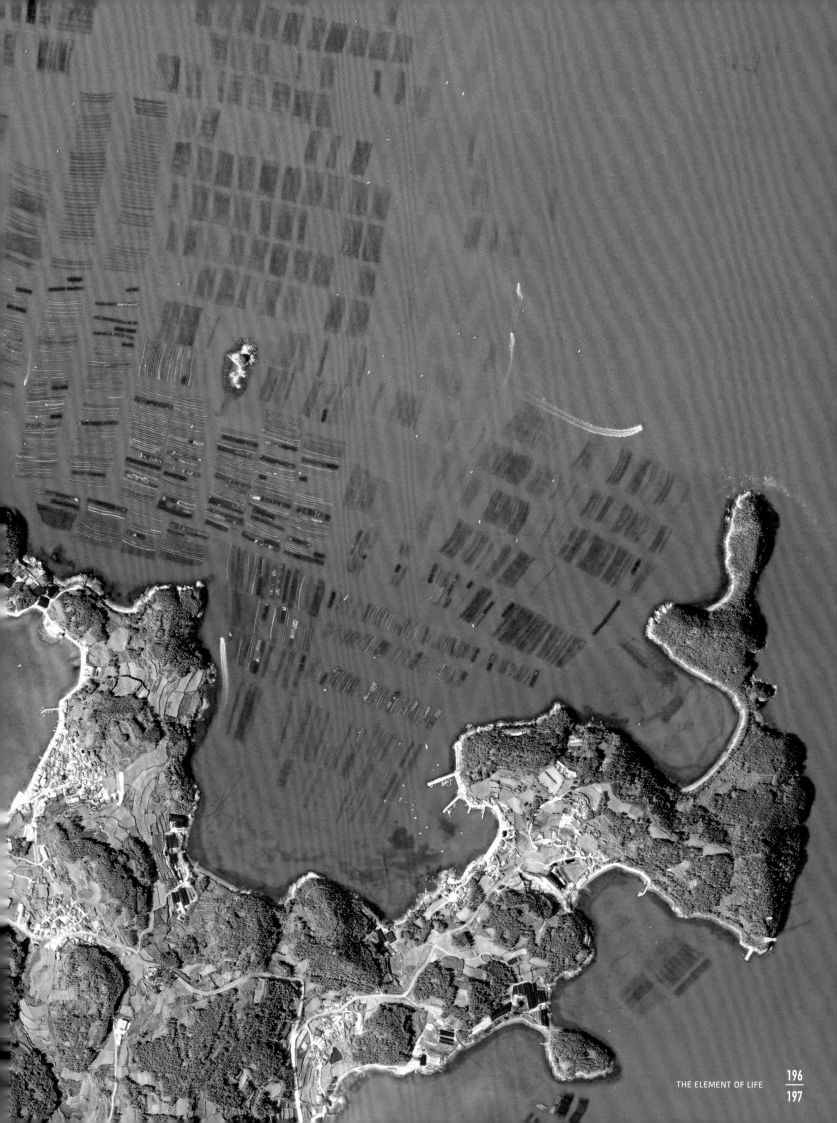

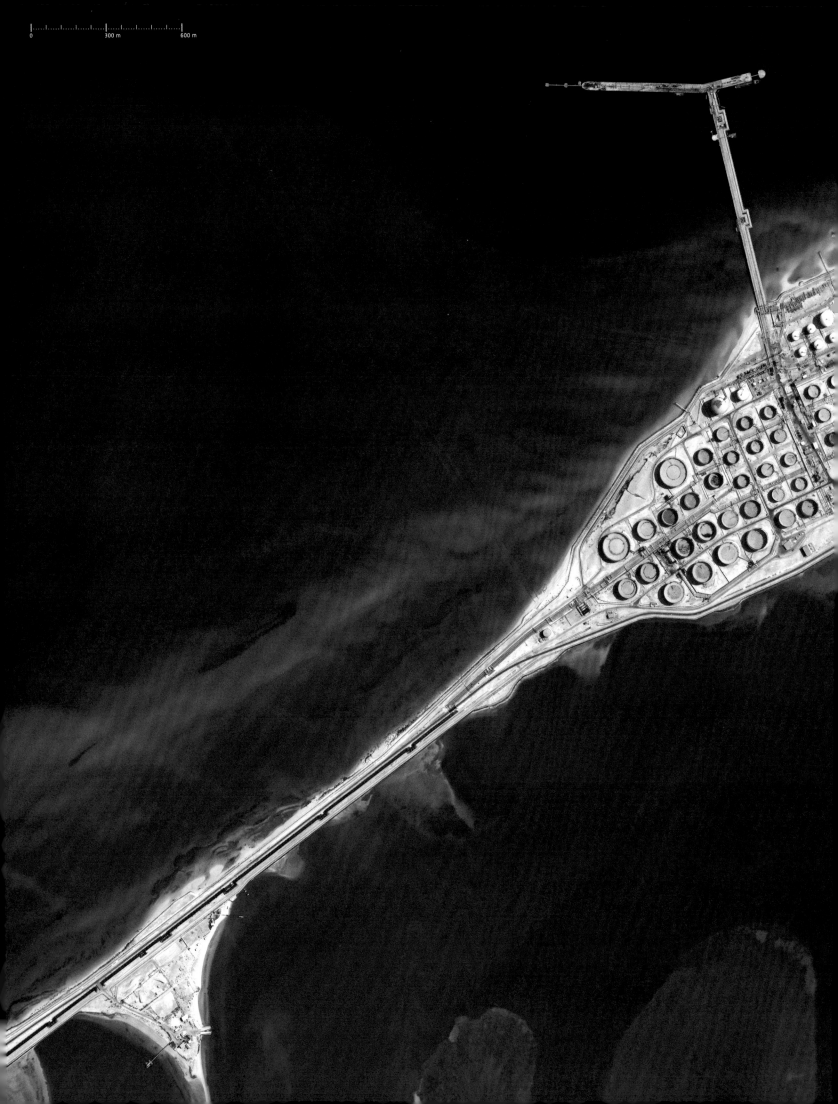

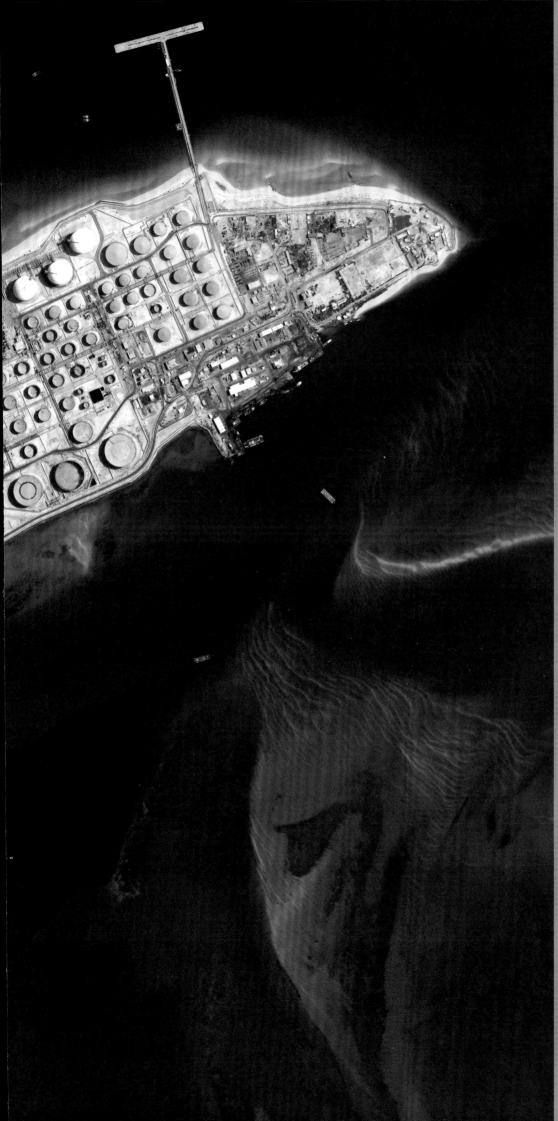

Utilized water
THE POWER
OF WATER

GAS STATION ON THE OCEAN: The peninsula Ras Tanura is
located near the Saudi Arabian Jubail; an oil refinery and the
largest oil loading port have been built here. At the piers and
on offshore pumping stations, up to 251 million gallons (950
million l) of mineral oil can be pumped into tankers every day
(26°39' N · 50°09' E; December 22, 2000).

ECONOMIC FACTOR: "WATER"

Anyone who watches children playing near water, whether at a mountain stream or at an ocean beach during a vacation, can often see how ships that were crafted from pieces of wood, float in the waters. You can experience how bridges and dams are being built, how moats for the defense of sand castles are created or, in the case of older children, how water wheels that have been stuck together with simple means are being driven by the current. This way, sometimes, a multitude of possible utilization areas of water become visible within a short time period for which water represents an important element for operating infrastructure facilities of humans.

The use of water for infrastructure facilities has in common that water is not consumed by them, but used, in order to subsequently be available in mostly unchanged quality for further use or to be returned to

nature. Important topics in this connection are water and transportation routes, as well as water and energy production; both areas are of extreme importance for life in our technology-based society.

Regardless of whether it is bridges, like the Golden Gate Bridge in San Francisco, canals like the Panama Canal or dam projects like the Three Gorges Dam at the Yangtze River, the large infrastructure projects constructed at water are among the most complex and spectacular structures built by humans. These structures emphasize the importance that water and its control has for modern society.

By river, by sea

With respect to traffic, water fulfils the role of a transportation route as well as that of a transportation obstacle. It forms natural barriers between neighboring populations, as well as efficient routes via which they can be connected to each other.

Originally, surely the separating effect of water was of foremost importance. Only after inventing aids like rafts and ships, did water surfaces become usable as transportation routes. However, as became clear in connection with the settlements and cities, the connecting as

"The sea has never been friendly to man.
At most it has been the accomplice of
human restlessness."

Joseph Conrad

← Multiple use

37°27' N · 38°44' E (August 14, 2013)

Dam lakes are important parts of infrastructure, which is also shown by the Atatürk Dam on the Euphrates River. Besides the power generation in the connected power plant with a nominal output of 2,400 megawatts, the dam lake is used as a reservoir for the irrigation of expanded areas in the surrounding region, especially southeast to the city of Urfa to which the water is guided through two tunnels.

0   125 m   250 m

well as the separating properties of water were used by humans from early on. Settlements and later, cities, were founded where natural ports facilitated easy access to the sea and where fords allowed for crossings of rivers and docking of ships. Before the development of efficient railroad systems and rural roads, rivers and the routes of coastal seafaring, were in many cases the only options to transport larger quantities of goods over great distances. Consequently, without exception, all the large trading centers were located at the sea or at navigable rivers where they represented the point of intersection between the traffic by water and by land. In many cases this still applies today. Venice, Genoa, Hamburg, London, New York or Shanghai, are just a few of these cities that have each developed into individual centers of trading networks.

Why cross the water?

Even today, there are risks associated with humans venturing out on water if they are not sufficiently prepared. Due to adverse weather conditions, icebergs or great depths, people are too quickly at the mercy of natural forces that they cannot counter with much. So there must have been very convincing reasons that made humans choose the route by water.

The actual reason for the central role of water for traffic is that it is the only substance on Earth that appears in noteworthy quantities in liquid form and is therefore mobile, as well as included in a permanent cycle. Following gravity, it makes its way in streams and rivers toward lakes and the sea, where it creates level surfaces that can only be disturbed by waves caused by the wind. Movement of water in rivers, or on the wide flat water surfaces of the oceans can be excellently used for energy-saving transportation by ships. Especially for bulk goods like mineral oil, coal and ore, shipping represents the most efficient means of transportation, in particular also because this freight is not perishable. Over the centuries, the oceans have therefore developed into vast natural traffic areas with high-traffic volumes for increasingly larger and more secure ships. Meanwhile, satellites help to increase the safety of ship traffic by determining the exact position of watercraft and being able to warn early of possible collisions, dangerous weather conditions or icebergs.

Highways on the water

Special focal points of sea traffic are created by the natural and artificial narrows, where the shipping routes that are otherwise distributed over wide areas are severely bundled. The coastal areas around such congestion points for sea traffic, for example at Gibraltar, the Bosporus or at the Suez Canal, are always also strategically important areas that are given special attention by political powers.

Natural shipping routes that were restricted by topographical conditions such as coasts, islands and shoals were, wherever possible, optimized by artificially created waterways. The construction of the Suez Canal, which connects the Mediterranean Sea with the Red Sea and thus the Indian Ocean, and the building of the Panama Canal, which created a navigable connection between the Atlantic and the Pacific Ocean, are some of the largest individual construction projects in human history. These two canals, their routes of which can be clearly identified in satellite images, have considerably contributed to shortening the trade routes. In both cases, they have not only drastically reduced travel times, but because of them, ocean areas that are particularly notorious for adverse wind and wave conditions can be avoided. Due to their enormous economic importance and the option to transport war ships on the canals, they also play an extremely important role in global politics.

↖ Ready for a long trip
1°16' N · 103°51' E (June 14, 2012)

The Tanjong Pagar terminal is just one of several terminals in the port of Singapore, which is the second-largest container port of the world behind Shanghai. Several hundred cranes load and unload the docked ships, moving an average of more than 81,000 containers in just one single day.

↑ Electricity for millions
25°25' S · 54°36' W (March 15, 2006)

The 20 generators of the Itaipú power plant turn the dammed energy of the Paraná River into up to 14 gigawatts of electrical power, with which the electricity demand of Brazil is covered by almost 17 percent and that of Paraguay by 75 percent. This makes the power plant second to the Three Gorges Dam as the hydroelectric power station with the largest output worldwide.

A safe harbor

Being the starting and destination points of shipping routes across the world, harbors play a special role. They often represent the central part of a city that has developed around them, and also carry the potential for conflicts due to the space requirements for storage and handling of goods. With respect to freight ships that are getting larger and larger, the biggest of which, with a size of approximately 1,310 feet (400 m) in length, and able to load more than 18,000 standard containers, efficient and reliable loading and unloading is vital for the proper functioning of a port. In some of the larger freight ports, container terminals are therefore becoming more and more automated. The two largest container ports are currently in Shanghai, where more than 32 million standard containers are handled annually, and in Singapore, with almost 30 million containers per year.

Parallel to the increase of the importance of ports for freight traffic, their previous importance as "Gate to the World" for passengers has considerably decreased because of the competition that aviation represents for long-distance passenger traffic, and ports are now almost entirely limited to acting as landing points for cruise ship tourism.

Tamed forces

As part of the discussions around climate change, to which the burning of fossil fuels like coal, mineral oil and natural gas contribute to the generation of electrical energy, hydropower has also increasingly moved into the focal point of interest. As a renewable energy form, it is considered one of the options that would use the resources of Earth in a more sustainable manner.

According to estimates from historians in China, the exploitation of the kinetic energy and potential energy of water reach back approximately 5,000 years, and were further developed over the millennia. But only the utilization of electricity as of the middle of the 19th century enabled deployment of the produced energy independently of its production location. Subsequently, the first large hydroelectric power stations were created before the turn of the 20th century, for example at the Niagara River in 1896. Currently, the total installed output of all hydroelectric power stations is at approximately 1,000 gigawatts and they create more than one-sixth of the total produced electrical power.

Energy landscapes

When viewing from space, the large hydroelectric power stations are generally clearly recognizable based on their dam walls and expanses of water in their backwater area. It also becomes evident that despite the image of hydropower as a sustainable energy form, they definitely have a significant impact on the environment. This becomes particularly clear at the dam projects on the large rivers of the world, such as the Three Gorges Dam at the Yangtze River, the Itaipú Dam at the Paraná River, or the Belo Monte power plant that is currently being constructed at the Rio Xingu River, an important feeder of the Amazon River. Such mega projects are often highly controversial due to their impact on the environment, as well as their negative social consequences for the residents.

Slower progress can be noted for the use of other forms of water energy. While tidal power plants in which the currents are utilized that are alternating between being directed toward land or the sea, depending on ebb and flow, have been in operation at suitable location for several decades already, the direct use of sea currents by means of underwater turbines, the same as the use of wave power, are still in the early stages of development.

Options for utilization and design

Transportation routes and the generation of electric power only represent two, although very important, utilization options of water in today's infrastructure networks. One of the features of water in particular, is its diversity, which allows its use also in many other areas. In many cases, the diverse utilization options result in multiple utilizations — for example, in the case of dam lakes that can be used simultaneously for the generation of power, as a reservoir for drinking water and for agricultural irrigation, as well as functioning as an element of flood protection for areas located downstream. Canals can also be used for water supply and as traffic areas at the same time. However, when the utilization becomes more intensive, multiple demands can also result in serious conflicts, for example when water that has been withdrawn for energy generation is no longer available for agricultural irrigation.

Waterways through the desert →
30°41' N · 31°53' E (March 31, 2014)

Approximately 75 miles (120 km) east of Cairo, the Suez Canal, which was built almost 150 years ago after a 10-year construction period, flows into the Red Sea. Through the Nile, the Suez Canal is connected with the Ismailia Canal, which branches off the Suez Canal north of the Bitter Lakes and reaches the Nile in Cairo. Because of the green band of irrigated agricultural areas, the route of the Ismailia Canal is clearly visible.

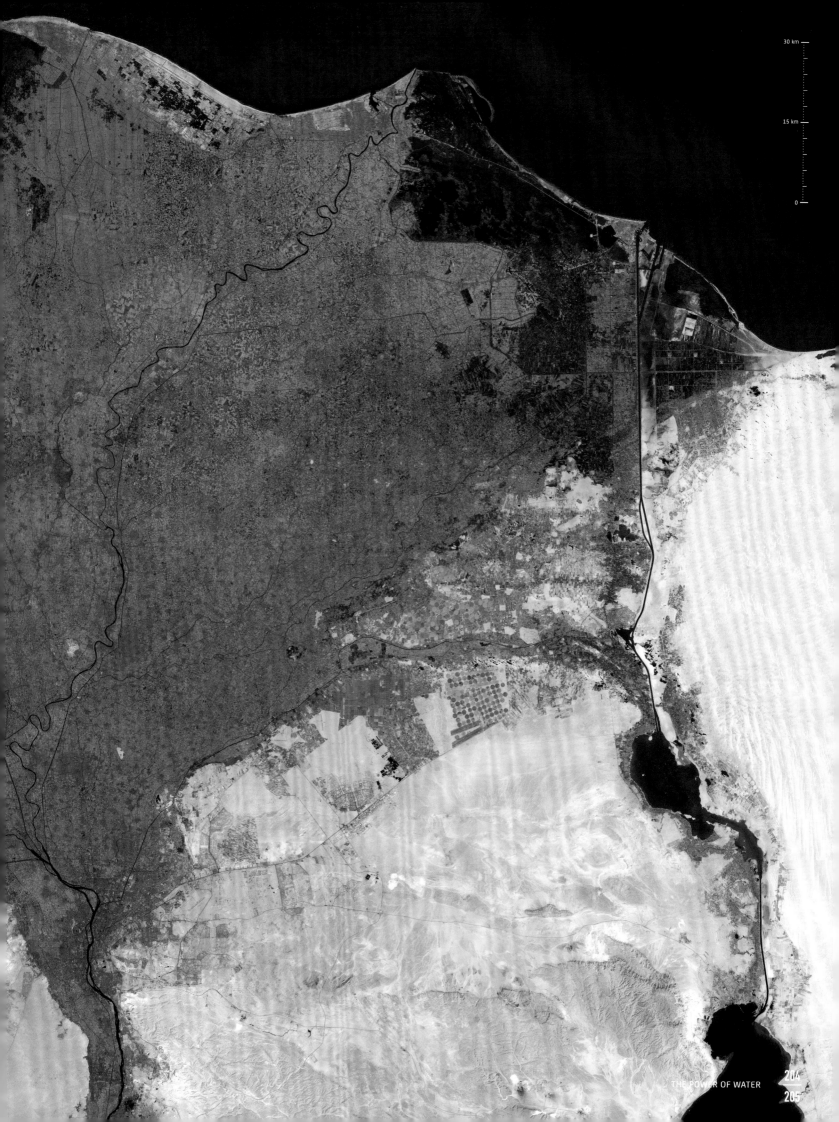

Center of the previous trade empire
44°24' N · 8°54' E (October 9, 2011)

In 2013, the harbor of Genoa, which was founded as
early as ancient times and later became the center
of a medieval trade republic, drew attention when a
freight ship rammed a control tower and made it collapse. Today Genoa is one of the most important trade
junctions in the Mediterranean region.

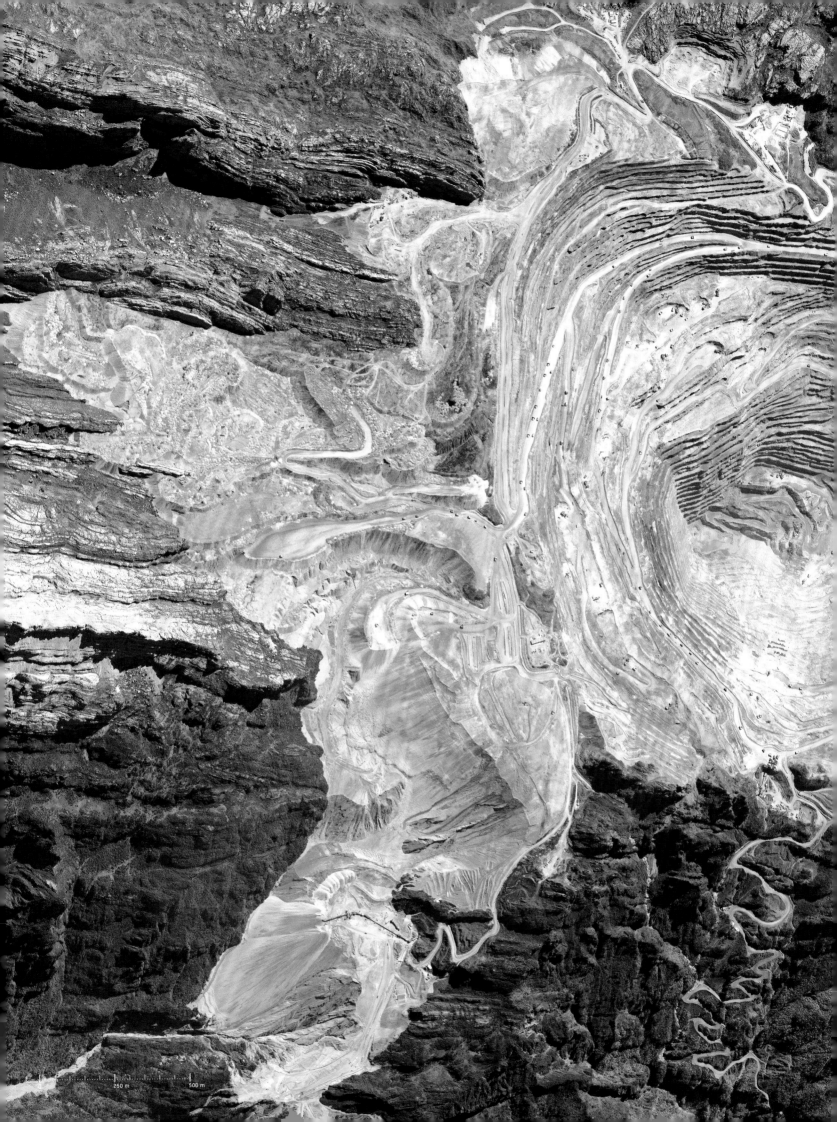

Gold rush in the jungle
4°03' S · 137°06' E (July 17, 2010)

Since the middle of the 1980s, gold and copper have been mined in the Grasberg mine at more than 13,120 feet (4,000 m) above sea level in the jungle of Papua New Guinea. The mine is considered the world's largest gold mine and third largest copper mine. Muddy waste water that is transported off via the Ajkwa River and is deposited along its course is an environmental burden.

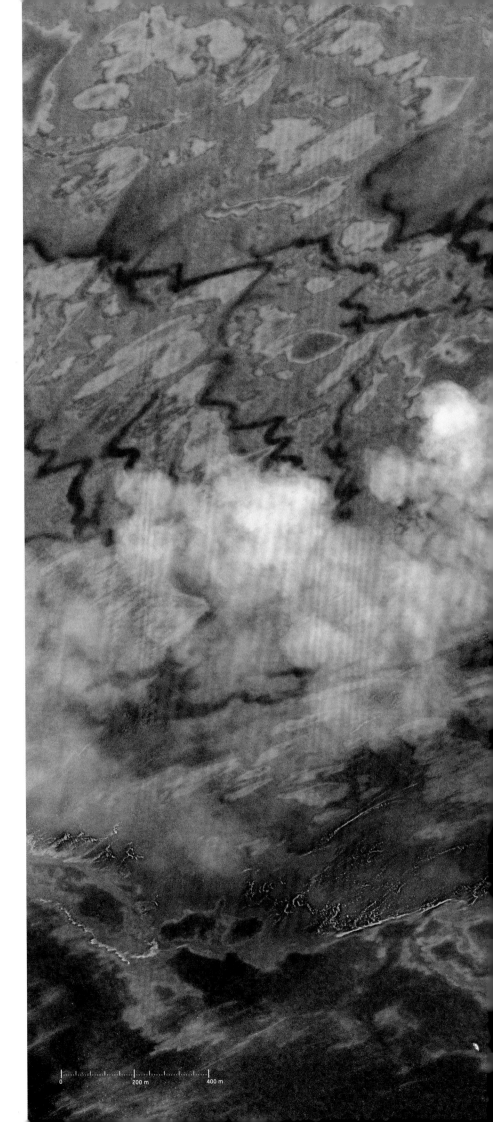

Oil painting on the ocean

28°57' N · 88°30' W (June 15, 2010)

The explosion of the offshore drilling rig Deep Water Horizon on April 20, 2010, released enormous quantities of crude oil, which caused extensive pollution mainly along the coast in the northeast of the Gulf of Mexico. Extraction by suction and chemical dissolving of the spill, as well as targeted burning, only resulted in partial success in the fight against the oil.

OIL DISASTERS

Disasters from oil tankers or oil drilling rigs with large quantities of mineral oil spillage are some of the environmental disasters that draw lots of attention in the media, since the oil floats on the water surface — clearly visibly after the accident. In addition, the environmental damages connected with the oil pollution, such as oil-covered seabirds and beaches blanketed in thick layers of tar, can hardly be concealed. This is why the perpetrators quickly introduce countermeasures after disasters, which however, often increase the damage rather than lower it. For example, in order to dissolve oil spills, chemicals are used, which are more harmful to the organisms living in the water than the oil itself. The cleaning of beaches with high-pressure cleaning devices can also have a negative impact on remaining creatures. As a result, one now relies more and more on the regeneration power of nature. Over longer periods of time, oil residue can also be degraded by natural bacteria.

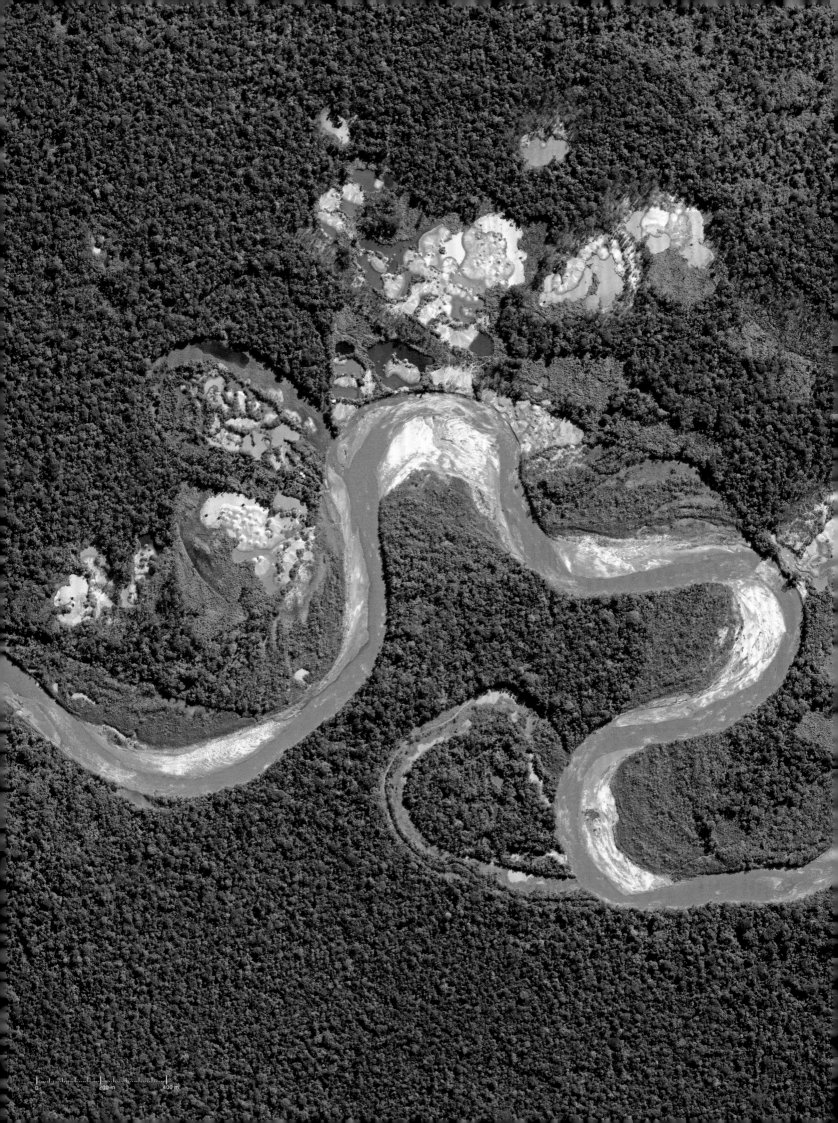

Rush of adventurers
13°01' S · 69°58' W (June 25, 2012)

During the conquest of South America, the search for gold always played a major role. This also applies to the Rio Tambopata in the Peruvian Southwest of the Amazon Basin, where gold diggers are usually digging illegally at the edge of the nature protection area Tambopata. Due to the gold rush, the population in the area has grown from fewer than 100 to over 30,000.

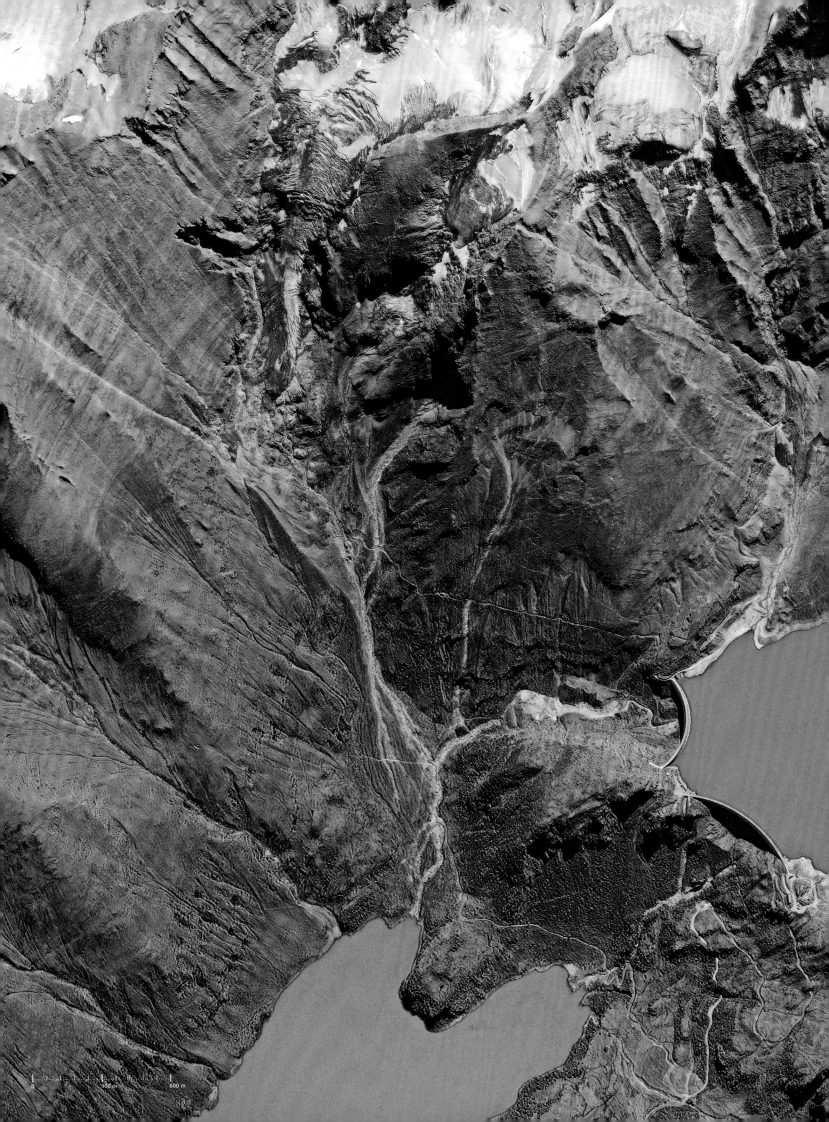

Energy on call

47°10' N · 12°44' E (August 22, 2011)

The dam lakes of the power plant group Kaprun are surrounded by the almost 12,465-foot (3,800 m) high peaks of the Glockner Group. The power plants use the difference in height to the Salzach Valley for the generation of electrical energy. The plants that were placed into operation in the 1950s, and were repeatedly modernized, can provide up to 810 megawatts of electricity. As pumped storage hydro power stations, they also serve as energy storage.

Direct disposal
21°00' N · 107°18' E (February 25, 2010)

The Vietnamese city of Campha is located at the Gulf of Tonkin directly between the sea and one of the largest coal mining areas in South Asia. Coal has been mined for over 100 years and together with fishery forms the economic base of the region. Waste water from the mine, amounting to approximately 318 million cubic feet (9 million cu. m) per year, is fed directly into the sea.

The color of the high-tech resource
23°39' S · 68°19' W (March 14, 2009)

Colorful evaporation basins cover large areas of the salt lake Salar de Atacama in Chile, which contains approximately 40 percent of lithium resources identified in the world. In the basins, the lithium carbonates are upgraded in the brine that has been pumped from the lake, until they can be gathered in the form of salt and be further processed.

IN THE LITHIUM RUSH

The fight for the resource of the future

In the Chilean lake Salar de Atacama, a substance is being mined that is needed in large quantities for the production of modern high-performance batteries: lithium. But mining for the precious metal from salt lakes possibly destroys the unique natural landscape.

For a long time, the resource lithium has lived a shadowy economic existence. Until the 1950s, lithium was almost exclusively used as an alloying element for the production of so-called white metal, which was used for parts on trains that are at risk of wear. Today, the range of applications reaches from the glass and ceramics industry, the production of high-performance batteries and the use in constructing aircraft and spacecraft, to the development of nuclear fusion reactors. Lithium has risen to become a treasured high-tech resource and its importance will increase even further against the backdrop of the announced development of electric mobility. Demand forecasts assume that in 2020, the requirements for lithium carbonate will be three to five times as high as today.

The rise of lithium to a modern high-tech commodity mainly has two reasons: First, the silver-white metal is uniquely light, and second, it has the highest standard electrode potential of all elements, which allows for the production of high-performance batteries.

The currently identified reserves amount to approximately 28 million tons. At a consumption of 22 pounds (10 kg) of lithium per electric vehicle, you could build 28 billion vehicles with this quantity. Since approximately 85 percent of the worldwide resources are located in Argentina, Bolivia, Chile and China, lithium could advance to an (economic) strategic resource in the future.

Today, only the resources in the salt lakes (salars) and the pegmatite deposits play a role for the commercial mining of lithium. Since the production of the mineral from the brine of salars is more economic than mining from pegmatite ores in mines, their economic importance is constantly increasing. Twelve of the 19 most important lithium deposits worldwide where lithium is currently being produced are brine deposits.

Behind the Salar de Uyuni in the Bolivian highlands, the Salar de Atacama holds the second largest lithium deposit identified to date. No brine of any other salar reflects a comparable level of lithium concentration. With 13,500 tons, more than 38 percent of the world production was produced in the Salar de Atacama in 2013.

The Salar de Atacama is located approximately 125 miles (200 km) east of the seaport of Antofagasta. It has a north-south expansion of approximately 55 miles (90 km) and a west-east expansion of close to 30 miles (50 km). With a total area of approximately 1,160 square miles (3,000 sq. km), it is the largest closed salt pan in Chile. The deepest point of the salar is approximately 7,545 feet (2,300 m) above sea level. The basin has had no drainage outlets for many millions of years and is additionally exposed to severe aridity. As a result, a massive body of evaporation and concentration rock has formed in the course of millions of years. The salar surface, which is mostly solid, consists of broad expanses of white table plane areas. Other areas reflect polygonal structures, and others again are characterized by bizarre towering salt formations. In some areas at the edge of the salar, you will find open lagoons with highly concentrated brines. Underneath the salt crust, with a thickness of around 20 inches (50 cm), the salar is similarly structured to a porous sponge, the shaped hollow spaces being filled with brine water, the so-called salmuera, up to an assumed depth of approximately 655 feet (200 m).

On one hand, water is supplied to the salar by means of (very low) precipitation and on the other hand, by a limited amount of inflowing ground and surface water. In addition to dissolved cations (like potassium or sodium) and anions (e.g., chlorides or sulphides), the water that flows into the salar includes lithium in the form of a minor element.

Mining the precious resource

The high lithium concentration in the Salar de Atacama was discovered rather coincidentally in 1963 by engineers from the mining company Anaconda, who ran a copper mine nearby. The first production facility in the Salar started operations in 1984 and has since been run by Sociedad Chilena del Litio (SCL). In 1994, a second mine with a considerably larger capacity was added, the Sociedad Química de Chile S. A. (SQM). In 2012 alone, the latter produced 48,000 tons of lithium carbonate and 6,000 tons of lithium hydroxide. By expanding its production facilities, SQM intends to increase its capacity to 68,000 tons of lithium carbonate in 2014.

Strong pumps are deployed to mine the lithium, which suck the salmuera from the body of the salar from a depth of 100 feet (30 m) and convey it into large evaporation basins. The salmuera, which reflects an average lithium content of 0.15 percent at the beginning of the mining process, now undergoes a concentration process by evaporation through the extremely strong solar irradiation and the wind. At a lithium content of 4 to 6 percent, the remaining yellow brine that contains mainly lithium chloride, is conveyed into a storage basin and subsequently transported by tanker trucks to La Negra, located approximately 12 miles (20 km) east of the seaport of Antofagast. Here, the final product, lithium carbonate, is produced from the concentrated brine.

Increasing threat to a unique habitat

The economic importance of lithium mining in the Salar de Atacama is undisputed. Therefore, the construction of new production facilities in eight other salars, which has already started in northern Chile, comes as no surprise. However, the effects of the lithium mines in the Salar de Atacama are devastating. At a consumption of several millions of cubic feet per year, it can be assumed that far more water is withdrawn from the salar than flows into it at the given climatic conditions. It is therefore feared that the ground water level drops and the open salt lagoons dry up because of the high amount of water withdrawal; this would in turn have a negative impact on the fragile desert flora and fauna.

Already, the extreme climate of the driest desert on Earth and the prevalent high salt content only provide a habitat for a few plants and animal species that have adjusted to these conditions. While different halophytes still grow at the edge of the salar, the central area is without vegetation. The lagoons are mainly used by three species of flamingos — the Chilean flamingo, Andean flamingo and the James flamingo — as a feeding and breeding area. The birds primarily live on small saltwater crabs, which themselves live on bacteria growing in brine water. The lagoons of the Salar de Atacama have been under nature protection since 1990; they are part of the national reservation Los Flamencos, comprising roughly 173,000 acres (70,000 hectares). Nevertheless, many of the flamingos have already moved away to the higher and more secluded small lagoons of the Altiplano. The increasing number of tourists — every year, several tens of thousands of visitors are registered — and the mine traffic in the salar, have probably driven the shy birds away. A drop of the ground water level by just a fraction of an inch, however, would not only result in the halophytes dying off at the edges of the salar, and would mean an end of the flamingo population and other species of birds. The destruction of this fragile habitat that has an irreplaceable ecological and cultural value for the local population would be sealed.

Professor Dr. Karl-Heinz Otto is head of the focus of work "Didactics in Geography" at the Geographic Institute at the Ruhr University of Bochum and is responsible for the Master Degree Program in the field of geography. In addition, he is chairman of the geographic Commission of Westphalia. Otto's scientific focus areas include the exploration of upland moors and the production of raw materials (primarily copper and lithium) in Latin America.

"A drop of the ground water level ... would result in the halophytes dying off and would mean an end of the flamingo population."

A wet airport
61°11' N · 149°58' W (May 24, 2013)

Anchorage is the economic center of Alaska and at the same time an important traffic junction for North America. Due to the topography, sea planes play an important role in the regional air traffic of Alaska. Therefore, a lake for the world's largest airport for sea planes was created directly next to the international airport. In the winter, starting and landing strips are set up on the lake ice.

ICY SEA ROUTES IN THE POLAR REGION

For centuries, the sea routes at the edge of the Arctic
Ocean being potential shortcuts of trade routes were the
destination for research expeditions. However, due to
the arctic sea ice cover, the Northwest Passage north of
Canada and Alaska, as well as the Northeast Passage north
of Siberia remained impassable for regular ships for long
periods of time. Only ice-breakers were able to break their
way through. However, due to the decline of the arctic
sea ice, the situation at the coasts of the polar sea has
changed. In 2008, the Northeast Passage as well as the
Northwest Passage were simultaneously ice-free for the
first time. Since then, the sea route through the polar sea
rapidly gains importance. In 2013, already more than 70
trade ships traveled through the Northeast Passage.

0   150 m   300 m

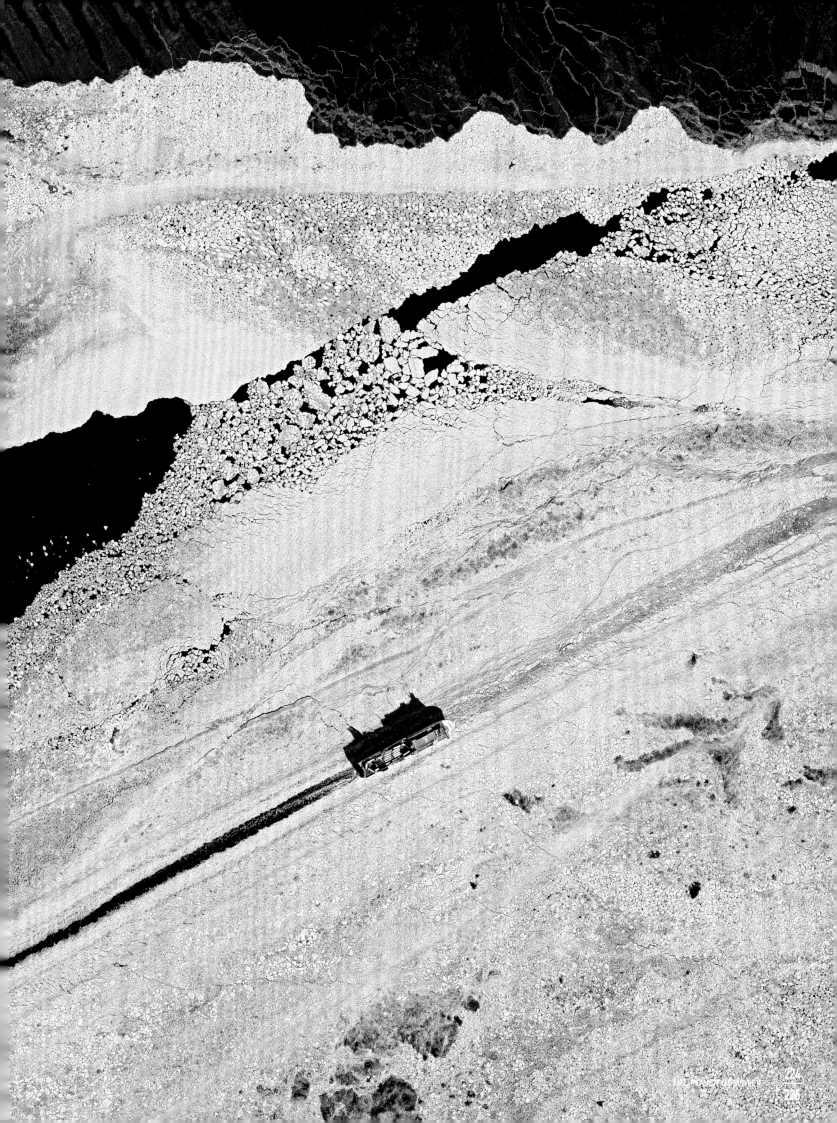

Sewage treatment plant for millions
41°49' N · 87°46' W (June 7, 2011)

The cleaning of waste water is one of the main important tasks of municipalities. In Chicago this is accomplished, among other things, by the largest wastewater treatment facility in the world, located 6 miles (10 km) west of the city in Stickney. Here, the waste water of almost 2.4 million people is treated in its catchment area. It is designed for a quantity of up to 159 million cubic feet (4.5 million cu. m) of waste water per day.

The dam wall of the Three Gorges Dam, which was taken into full operations in 2008, holds the Yangtze River back over 375 miles (600 km). In addition to producing more than 80 terawatt hours of electric power annually, the dam is also intended to improve the flood water protection and navigability of the river. The economic gain is confronted with serious ecological and social problems for the original residents of the area that is now flooded.

HYDROPOWER — IN HARMONY WITH NATURE?

Similar to wind energy, solar energy and biomass, the generation of electricity by utilizing the kinetic energy of water is generally considered an environmentally-friendly and sustainable method. However, also here there are impacts that must not be neglected. The construction of dams alone is often linked to deep intrusions into the surrounding area, the extent of which increases as soon as the backwater process through the power plant begins. Besides the loss of natural habitat for many species of animals and plants, the flooding of the home of many people who must be relocated in the course of the construction work is often accepted. These negative impacts are especially distinctive for large projects like the Three Gorges Dam at the Yangtze River in China or the Brazilian Rio Xingu Project, where rivers are held back over hundreds of miles.

Main artery of freight traffic
29°57' N · 32°34' E (March 2, 2011)

The Suez Canal spares ships the lengthy trip around Africa. Since it leads through an area with only slight elevations of the terrain, sluices are not necessary. The first canals between the mouth of the Nile and the Red Sea were therefore already built in ancient times. Opened in 1869, today's Suez Canal represents an enormous relief for the trade with Asia and East Africa, and therefore plays an important geopolitical role.

0	300 m	600 m

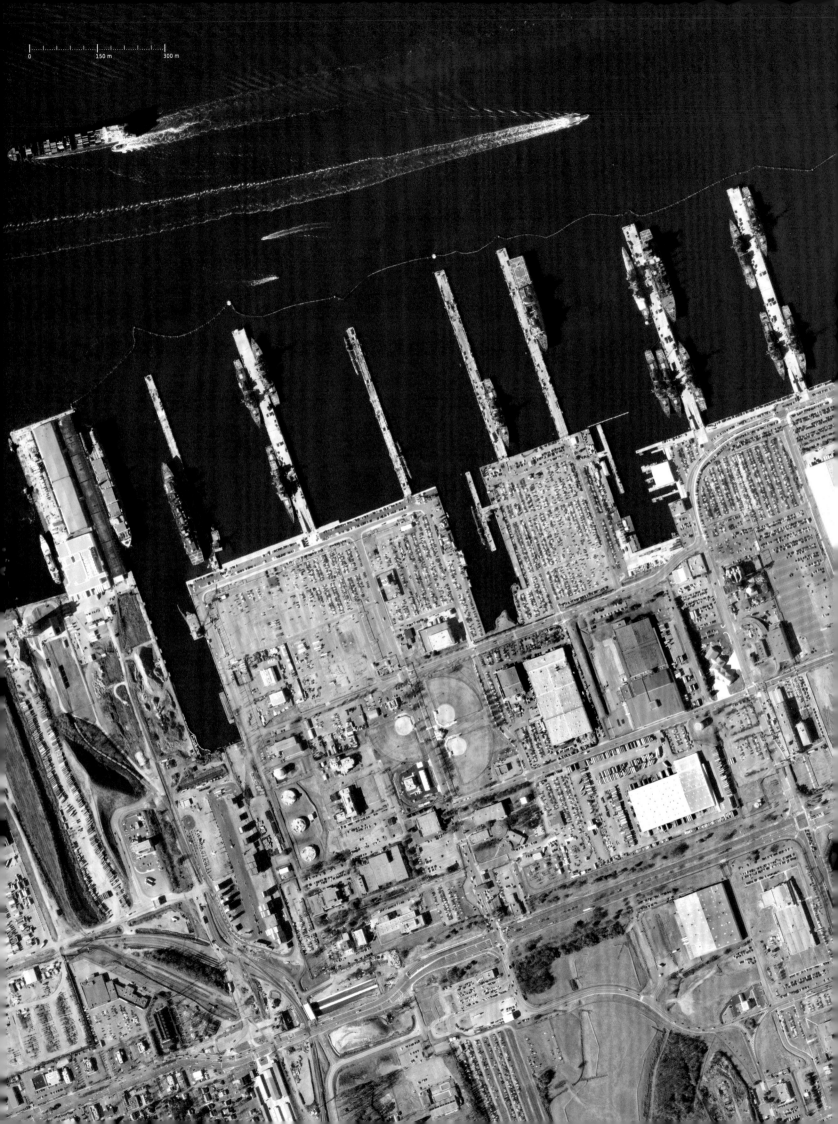

Floating military airports

36°57' N · 76°20' W (December 14, 2012)

The world's largest naval port is located in Norfolk,
280 miles (450 km) south of New York at the Atlantic
Ocean. The base is the home port for many war ships,
including some of the largest aircraft carriers and
cruisers of the U.S. navy. The expansion of the parking
lots in front of the piers demonstrates the crew sizes
of these ships.

End of a raft trip
49°15' N · 123°15' W (April 23, 2009)

Rivers have always been used as energy-saving transportation routes for wood. For rafting, tree trunks are bound together and float downriver with the current. This is also carried out in the vast forest areas of the Canadian province British Columbia, where the wood finally arrives at the mouth of the Fraser River where it awaits further transportation.

Container over container
1°16' N · 103°50' E (June 14, 2012)

Due to its limited space, the island state of Singapore has the third highest population density of all countries behind Macau and Monaco. A large portion of the area is furthermore claimed by the logistics areas of the port facilities, which Singapore's role in the world economy are founded upon. In 2013, more than 32 million standard containers were handled, which makes Singapore the largest port in the world.

ECONOMY ON A TRIP AROUND THE WORLD

In the middle of the 1950s, a new solution was found in the form of standard containers for handling the increased worldwide freight transportation. With standardized lengths of 20 and 40 feet (approximately 6 and 12 m), these containers hold a load between 21 and 40 tons. In addition, they can be stacked without gaps and reloaded from a ship to a train or a truck, regardless of their content. Thereby, they constitute the prerequisite for efficient freight handling and for comparatively low transportation costs and thus also one of the foundations of the globalized economy. The approximately 30 million standard containers that are handled in the second largest container port of the world, in Singapore, alone, demonstrate the extent of the transportation of cargo worldwide. In view of these figures, the 2,000 to 10,000 containers that are lost at high sea, according to insurance companies, hardly carry any weight.

Ship traffic without junctions
52°18' N · 8°56' E (May 1, 2011)

In the German city of Minden, the Midland Canal
crosses the Weser River. Since 1915, a single bridge
was used to cross the river; in 1998, a second bridge
was added. As a result, not only were the two bodies
of water decoupled, but smooth ship traffic was also
ensured. Ships can switch the waterways by means of
sluice systems.

Building material from the river
23°48' N · 90°19' E (November 29, 2008)

Every year during the monsoon season, the rivers around Bangladesh's capital of Dhaka are in flood and large areas around the city are flooded. At a tributary of the Buriganga River, an entire branch of industry is based on the residues from the floods. Mud is collected in numerous small brickyards and processed into bricks.

37°25' N · 141°02' E (August 19, 2013)

Resulting from the tsunami on March 11, 2011,
the supply with cooling water failed in the nuclear
power plant Fukushima Daiichi. In the following
days, meltdowns occurred in three of the four reactor blocks; large quantities of radioactive materials
were released. Tens of thousands of people had to
be evacuated. Since then, contaminated cooling
water is collected and stored temporarily in numerous tanks on the property of the power plant.

POWER PLANT AS A HEATING
DEVICE FOR THE ENVIRONMENT

In the context of power generation, water is almost exclusively viewed as an energy provider in hydropower plants.
But also many other power plant types are dependent
on water, which is used as cooling water or a means for
energy transfer, e.g., in the case of steam turbines. The
water quantities that are used for cooling thermal power
stations are enormous, regardless of whether they are
operated with coal, gas or nuclear power. Consequently,
high-performance nuclear power plants are usually built
at bodies of water from which sufficient cooling water can
be withdrawn. A failure-safe construction of the cooling
systems is particularly important since their failure can
lead to catastrophic nuclear accidents.

Limited space to live

4°11' N · 73°26' E (February 11, 2013)

The artificial island of Thilafushi is located just a
few miles off Malé, the capital of the Maldive
Islands. Waste from the island state is deposited
here and covered up with layers of construction
debris and sand. Since the start of this work in 1992,
Thilafushi has grown to almost 5,000 square feet
(500 sq. m) and is now also used as building land for
industrial companies.

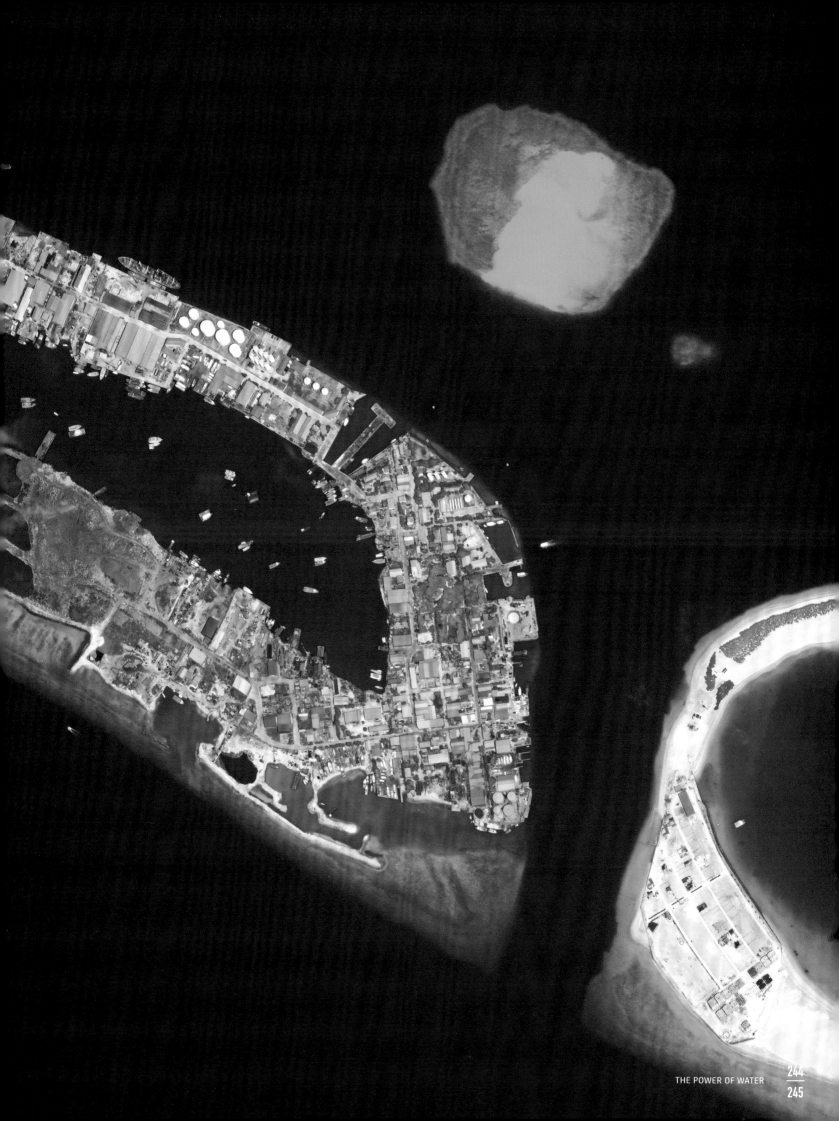

Treasured shortcut

9°00' N · 79°36' W (February 1, 2013)

Its construction was already started in 1881 — an
indicator of the importance of the shortening of the
ship route around the southern tip of South America
was even back then. In 1914, the Panama Canal was
finally opened. The capacity of the approximately
50-mile (80 km) long waterway has been increased
again and again. Every year almost 15,000 ships,
constituting approximately 5 percent of all sea freight
worldwide, pass through it.

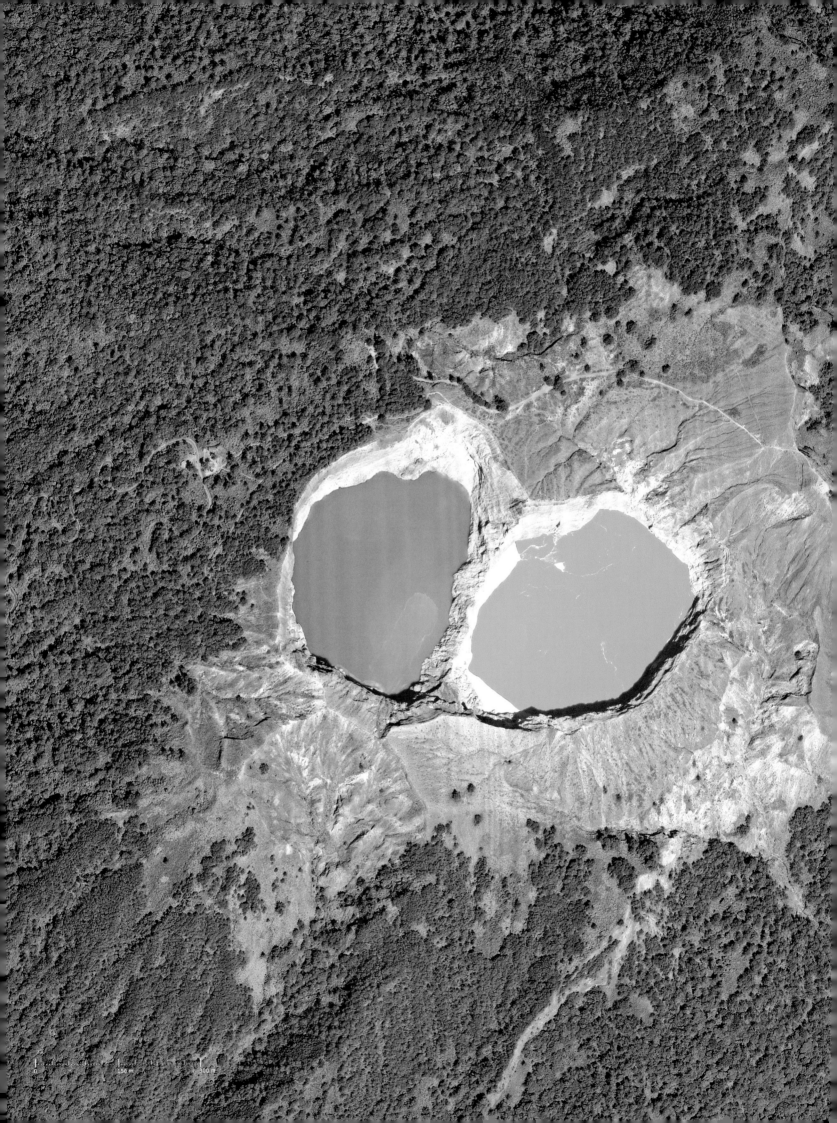

PARADISES ON THE WATER

FROM THE COLOR PALETTE OF NATURE: Three lakes fill the crater of the Kelimutu volcano in the Kelimutu National Park on the Indonesian island Flores. The crater lakes reflect different colors, which change depending on the time and are probably caused by chemical reactions due to volcanic activities (8°46' S · 121°49' E; September 2, 2011).

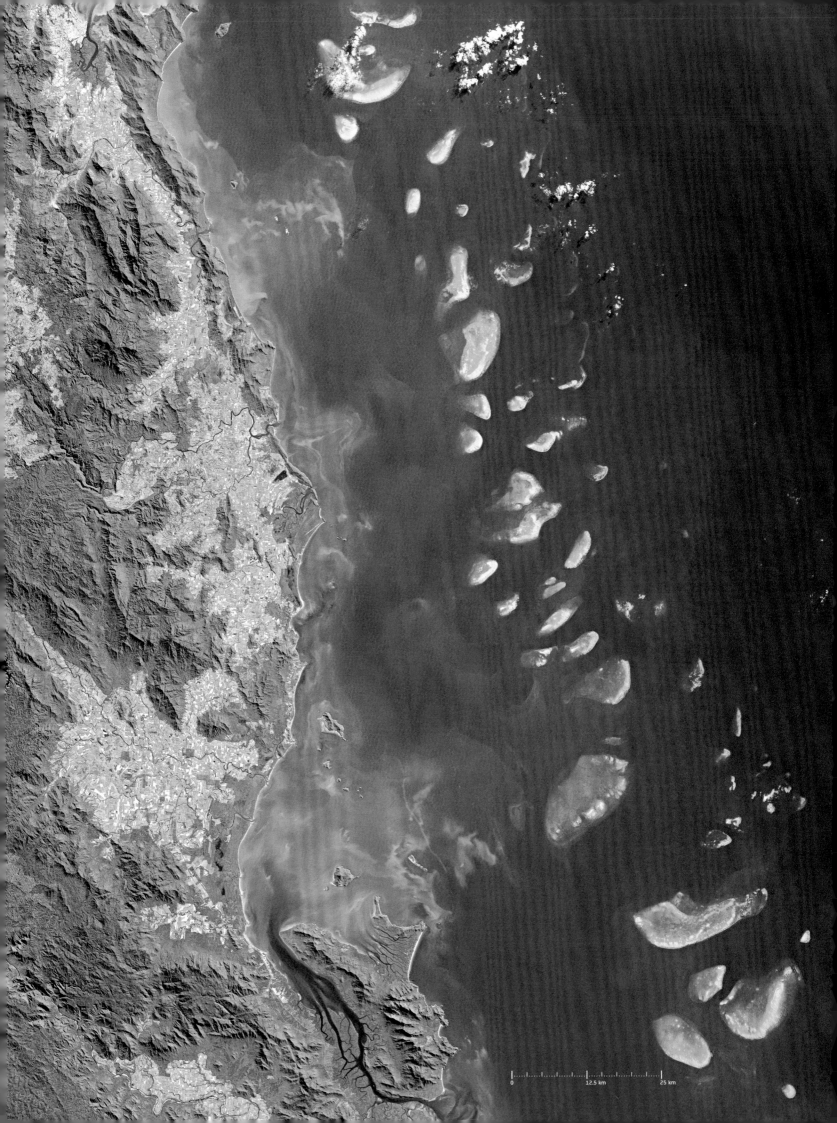

Water and nature
UNSPOILED LANDSCAPES

I n 1979, the American submarine "Alvin" went on expeditions to explore the ocean bed above the Juan de Fuca Ridge off the coast of the northwest Americas. The hope was to make new findings about the processes of plate tectonics here, where the meeting of continental plates causes distortions and faults in the continental crust. The scientists were soon successful and discovered so-called "black smokers," chimney-like hydrothermal springs that release hot water with mineral content from the ocean floor.

However, what they found in this environment was a sensation. The ocean floor around the vents of the black smokers was swarmed with previously unknown life forms that were not only able to survive a pressure that is up to 250 times as high as atmospheric pressure and temperatures above 750°F (400°C) without sunshine, but actually needed these conditions to live.

As this example shows, life on Earth has occupied many niches that often appear hostile to life at first glance. From the polar ice worlds and dried up dune landscapes of the large deserts, to deepest parts of the oceans, specialized life forms can be found everywhere. Having adjusted to often extreme environmental conditions, they have conquered habitats by using their special attributes.

However, despite all the differences between the habitats on Earth, they have at least two attributes in common: energy sources and water in some form must exist, even if in the smallest of quantities. If an environment fulfils these conditions, it is almost a certainty that life can be found — and naturally, especially the habitats directly in and at water, in the rivers, lakes, oceans and wetlands are particularly favored.

As diverse as the ambient conditions on Earth may be, as varied are also the habitats and life forms to be found. The most different types of bacteria, algae, fungi, plants and species of animals live in the so-called biographical zones of the planet and each form their own biocoenosis in which they are dependent on each other in a dynamic balance. Completely natural habitats, still in their original

"Everything originated in the water!
Everything is sustained by water!" Johann Wolfgang von Goethe

← Australia's pearl chain
17°41' S · 146°16' E (July 5, 2002)

The bright reefs and islands of the Great Barrier Reef lie off the northeastern coast of Australia like a necklace. With a length of over 1,430 miles (2,300 km) and an area of almost 135,135 square miles (350,000 sq. km), it is the largest reef worldwide. Although most of the reef has been placed under nature protection, it is exposed to different threats, from which climate change has the severest impacts.

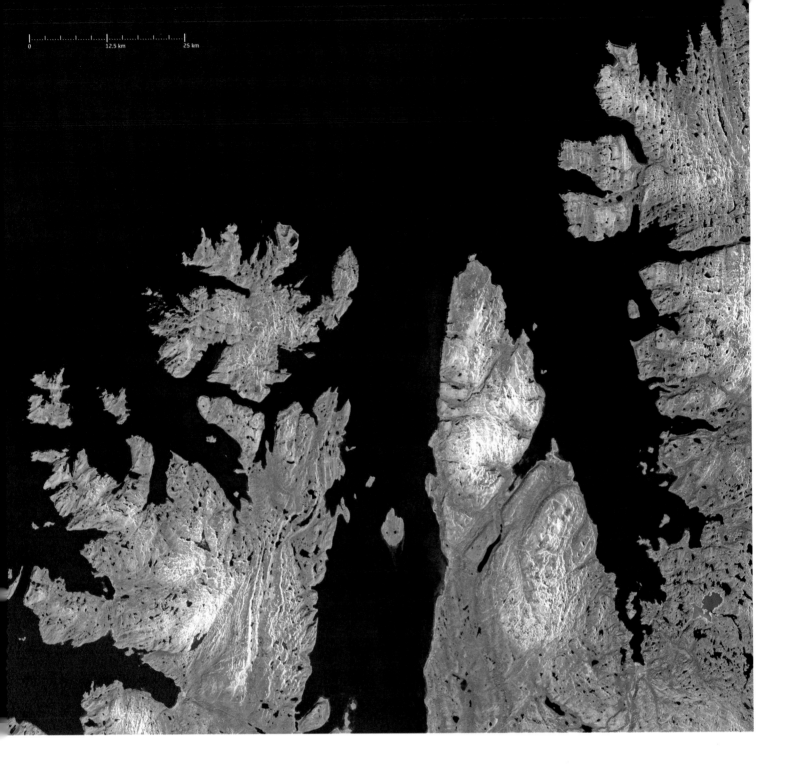

0   12.5 km   25 km

balance, have become rare due to the expansion of humans. All the more important it is to protect the remainders of natural landscapes and save them as refuge for numerous species of animals and plants.

The aforementioned example of the detection of the black smokers is just one example of how diverse and adaptable life on Earth is. In an era in which space probes and robots explore our neighboring planets, this shows how unknown our home planet still is in many areas, and how important it is to keep changes as low as possible when it comes to environment-changing activities.

About rivers and lakes

Clear mountain lakes, babbling brooks rushing over rock and peaceful meadowy countryside — the familiar habitats in connection

with water are often linked to such images and are mainly formed of streams and rivers, and standing waters of the ponds and lakes. Depending on the respective climate conditions, these habitats are shaped very differently. Mountain lakes fed by glaciers, which only have a few animals living in them because of the low water temperature and low nutrient content, are seen alongside bodies of water that interact with dense vegetation at the water's edge and that can offer a home to a multitude of different species of animals and plants with their associated nutrient range.

However, especially in the surroundings of rivers and lakes, natural habitats have become rare because the diverse utilization possibilities for humans have led to strong competition with nature. A large portion of the rivers have become regulated in order to secure settlement

areas and cultivated land, but also to provide the residents with protection against floods. With this development, many meadowy countrysides, which represented important habitats because of their complex system of flowing and standing waters, disappeared. Several moors and wetlands were drained for the expansion of agricultural areas. In addition, many rivers were dammed for the generation of electric power and for water storage which had severe impacts on the migration possibilities of fish, for example. Fishing and discharging of waste water represent additional adverse effects on the habitats in rivers and lakes.

One known example for the consequences of intrusions by humans is in the east African Lake Victoria. Intensive agriculture in its

↑ At the northern edge of Europe
70°57' N · 25°54' E (June 21, 2001 and October 27, 1984)
Although it is one of the areas of the European mainland that reaches furthest North, the North Cape remains ice-free even during the winter; the reason is the warmer water that is passed from the extensions of the Gulf Stream around the northern tip of Scandinavia.

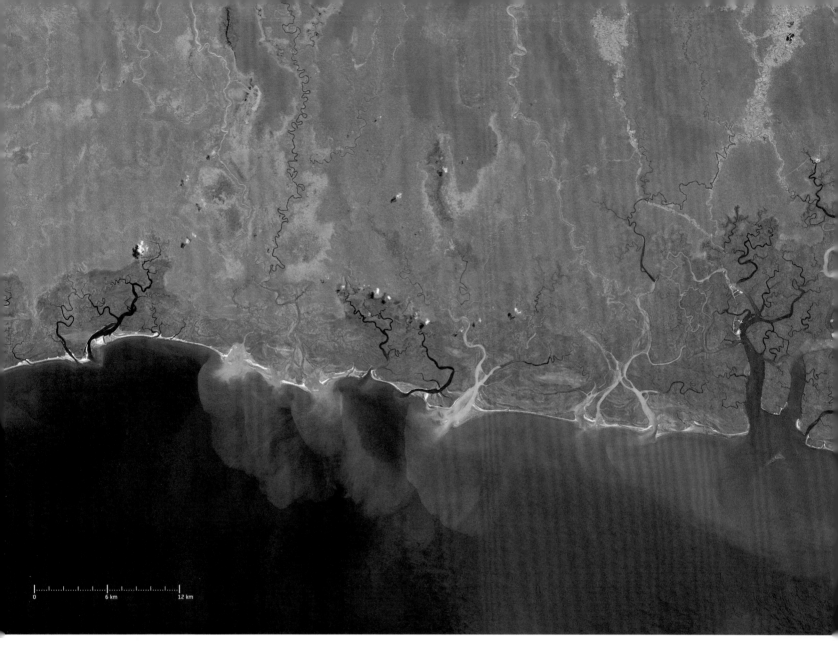

0   6 km   12 km

surroundings and the discharge of waste water have resulted in an increased supply of nutrients. This caused the water hyacinth, which was originally not a native plant to grow explosively. This development was encouraged by the fact that the Nile perch, which was introduced for fishing, had severely decimated the native plant-eating species of fish.

Today, natural bodies of water are mainly located where the interest of man for economic activities is too low. The reason is either return that is too low, or too much effort to realize this return. Therefore, today, the largest percentage of natural areas can be found in areas that are difficult to reach with unfavorable climate conditions for humans. Examples are the icy areas of the tundra lakes at the edge of the Arctic and the river systems in the rain forest areas of Amazonia and Central Africa. But even here, the utilization is progressing, whether due to the discovery of mineral oil in the Arctic, or because of road building through rain forest which opens previously unreachable areas.

The wealth of the oceans

Oceans cover 70 percent of the surface of Earth and reach to a depth of up to 36,000 feet (11,000 m). They contain more than 96 percent of the water that is available on Earth. The importance of the oceans for the climate cannot be valued highly enough. The surfaces of the oceans are important sources for rising vapor, which in turn results in the formation of clouds and precipitation. Ocean currents contribute to the redistribution of irradiated sun energy and therefore reduce the temperature differences on Earth. Life in the oceans also influences the composition of the atmosphere. For example, the plants in the oceans are responsible for over two-thirds of the global production of oxygen.

Due to their size alone and the fact that they stretch over all climate zones on Earth, oceans provide space for an incredible number of the most different habitats, which interact with each other in complex ways that are often not yet understood. And the inhabitants of these spaces are just as diverse as the habitats. Bizarre creatures of the deep sea that often appear almost transparent, but can also

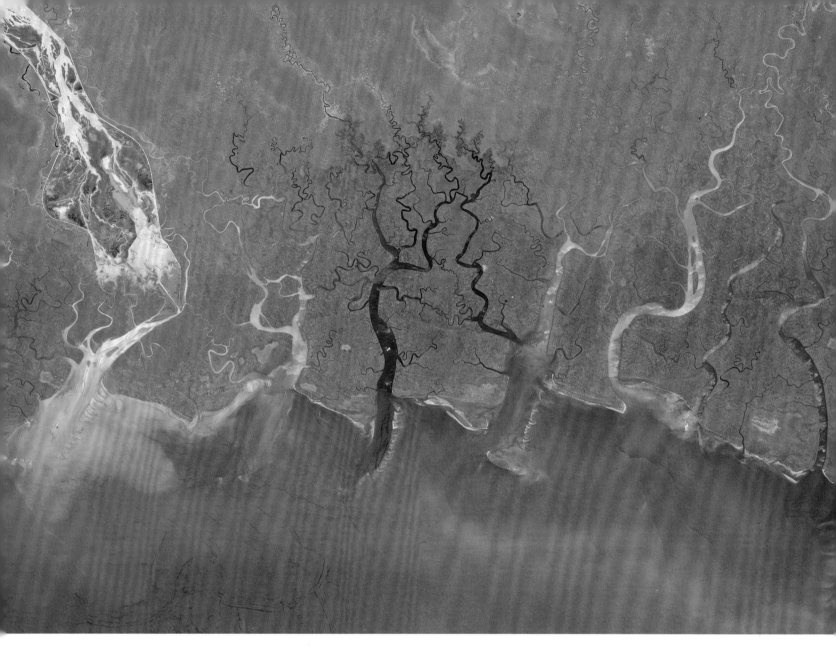

glow in order to attract prey, always cause amazement and serve as sources of inspiration for designing alien creatures in science fiction movies due to their strange appearance. The largest species of fish, the whale shark can reach a length of more than 40 feet (12 m), and the largest mammal on Earth, the blue whale, which can reach a length of up to 108 feet (33 m) and a weight of 200 tons, also lives in the sea. Divers love the coral reefs of the tropical seas, which symbolize almost ideally the power of life on Earth with their colorfulness and large number of the most different species of fish.

The island worlds represent special habitats that are closely connected with the oceans. Having been cut off from mainland for a long time, in many cases they have brought forth unique creatures that have ideally adjusted to the living conditions of their environment and can only be found in these surroundings. These endemic species include the marine iguana on the Galapagos Islands, the Kiwis in New Zealand, but also an extinct species of human (which was only discovered in 2004), who were of short stature and called "hobbits" — 18,000-year-old fossils of which were found on the Indonesian island of Flores.

↑ Tropical swamps
4°48' S · 136°52' E (April 1, 2014)

The coastal area in the Indonesian part of the Island of New Guinea appears almost untouched over vast areas. Mangroves and marshy landscapes with meandering rivers shape the scenery. In the east, you will find extensions of the Lorentz National Park, which is one of the areas of the island with the greatest diversity of species. During the rainy season, high sediment discharge of the rivers colors the mouth areas into the Arafura Sea.

Limited sizes

Since their vastness appeared immeasurable to humans for a long time, man has only started to think about the limits of oceans and their resources a few decades ago. This limit is mainly noticeable in connection with fishery, the enormously grown effectiveness of which has resulted in overfishing of most fish stock and in some cases, already to a threat to their continued existence. Another specific threat, from which nature can only recover over long periods of time, is the widespread contamination of ocean areas due to ship disasters, especially oil tankers and accidents on offshore oil production facilities. The impact of removing mangrove forests at the coasts of tropical oceans should not be underestimated when these forests are replaced by aquacultures, for example. Setting up protected areas is an important measure in order to preserve at least some of the ocean areas that are the most important for the diversity of species to a large extent in their natural state.

Indirect effects that are associated with the climate change are added to the direct intrusions into the marine ecosystems. Warming of the seawater, rise of the sea level and acidification of the oceans due to an increased absorption of carbon dioxide from the atmosphere, contribute to the environmental conditions of the oceans changing. Impacts can already be observed in the coral reefs, for example in the Australian Great Barrier Reef, which is one of the largest ocean protection zones in the world and yet already shows severe damage.

Icy deserts

Similarly to the deep sea, the icy worlds of the ice caps and glaciers are extreme habitats in which only specifically adjusted species of animals and plants can live. On land, higher plants are not relevant, since they cannot survive on ice forming the ground. Being mobile creatures, animals are far better suited to handle the extreme conditions in these surroundings. Even so, in comparison to the vast expanse of the Antarctic, which is larger than Europe, the number of species of animals and of individual creatures is small.

Emperor penguins are an excellent example of adjustment to life in the hostile environment at the edge of the Antarctic. They defy the constant cold, as well as darkness of the polar night that lasts for months, and even manage to go without food during the approximately two-month-long breeding season. By means of satellite images, the population of this species was estimated to almost 600,000 animals.

Icy environments in small spaces can be found in the regions of mountain glaciers on all continents, with the exception of Australia. In conjunction with the backdrop of the mountains, which is often quite impressive, they are part of the standard inventory of alpine tourism. However, the glaciers that are usually located at higher elevations also represent very special habitats, each with their own microclimate. In a large number of protected areas, an attempt is made to protect these special habitats against intrusions by humans. The impacts of climate change, in particular the warming of the Earth's atmosphere, are also noticeable here. A large portion of the glaciers — especially in the northern hemisphere — is retreating, which causes a rise of the sea level in addition to the loss of these habitats.

Protection of habitats

Natural habitats became rare and not only in connection with water. The main reason is the intensive utilization of Earth by humans, which extends increasingly to more regions of the planet and meanwhile also includes the icy areas of the polar regions.

More importantly are the national and international efforts to protect the remaining natural habitats. Programs such as the World Heritage List of the UNESCO, where cultural heritage is included besides natural heritage, play an important role in promoting awareness for the importance of preserving these areas. More specific goals are pursued by the Ramsar Convention for example, the focus of which is the protection of wetlands, or the World Conservation Union IUCN, which contributes to an international comparability of protection efforts by creating Red Lists of endangered species and categorizing protection areas.

Against the backdrop of the global developments of past decades, the growing understanding of our Earth as an overall system from which we humans live and which we redesign to an increasing extent, slowly leads to rethinking. The necessity to rethink the general handling of the Earth's resources, by exceeding local to regional attempts to protect nature, so that our home planet can be preserved in its amazing diversity, is becoming increasingly evident.

Brown caiman and giant guinea pig →
17°52' S · 57°27' W (June 16, 2011)

In many regions of Earth, precipitation varies considerably depending on the season. This also applies to the South American wetland area of the Pantanal, which is flooded over large areas during the rainy season. The region is home to many animals, including 6-foot (2 m) long giant otters, capybara who are relatives to guinea pigs and brown caimans.

0   250 m   500 m

Mysterious desert lakes

40°02' N · 102°12' E (May 6, 2010)

More than 140 small lakes are located between the
dunes of the Badain Jaran Desert in China's Inner Mon-
golia. They also give the desert its name, which means
"mysterious lakes," translated from Mongolian. Partly
fed by groundwater springs, the lakes have greatly
varying salt contents.

Island paradise for seabirds

53°43' N · 8°08' E (May 13, 2009)

The island of Mellum in the "National Park Nieder-sächsisches Wattenmeer" (Lower Saxon Wadden Sea National Park) only emerged around the turn of the 20th century as a result of ebb and flow. Since then, the dune island has grown to an area of 1.7 square miles (4.5 sq. km). Vast sand and tideland areas have made Mellum an important breeding area for seabirds and one of the most important habitats for seals in the North Sea.

0   250 m   500 m

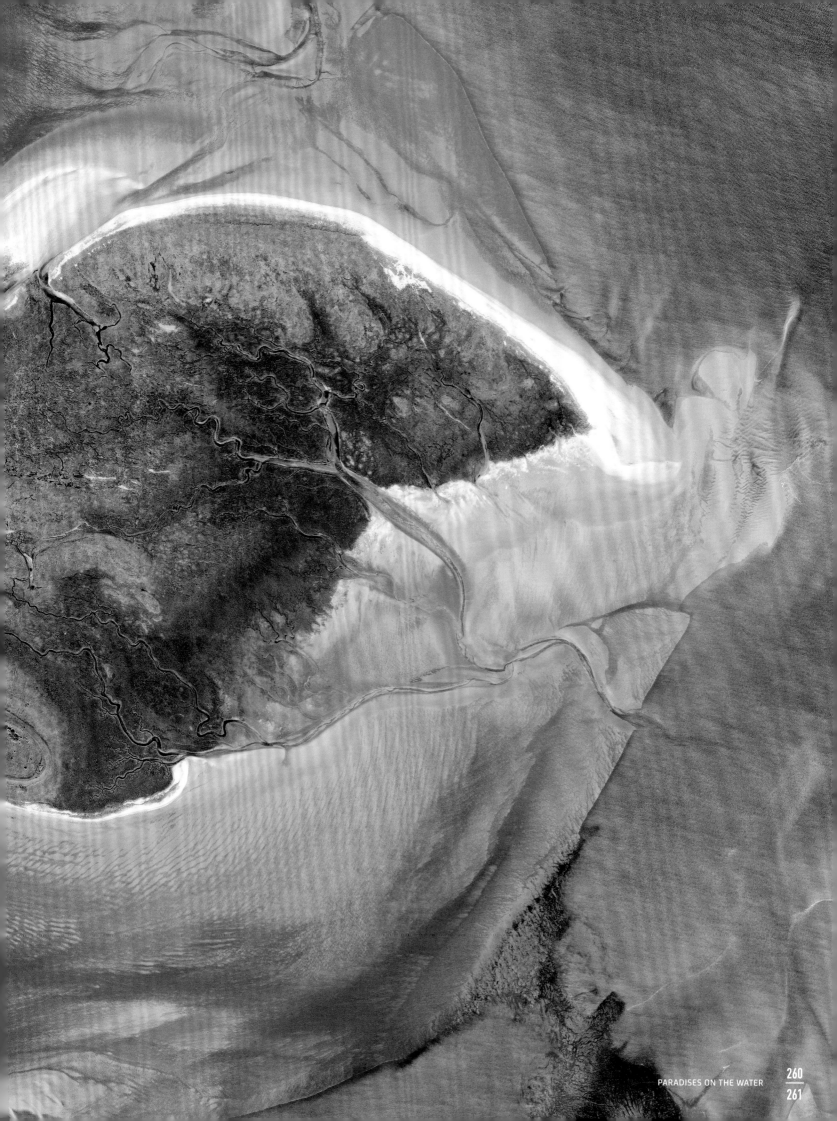

Icy borderlands

71°42' S · 76°56' W (November 19, 2002)

In the Antarctic summer, the sea ice that formed in the winter, dissipates and floats in sheets of ice on the ocean. The anchor-shaped Smyley Island is only a few miles away from Alexander Island off the Antarctic Peninsula. The area became known at the beginning of the 21st century because the collapse of the Wilkins Ice Shelf could be observed here.

WHITE, WIDE ENVIRONMENTS

Until recently, the extreme habitats of the Arctic and Antarctica were safe from massive intrusions by humans. The effort opposing more extensive activities in this hostile environment was too high. Now, two developments contribute to slowly changing this situation: First, the technical possibilities are increasingly better suited to cope with the icy conditions and second, climate change results in the fact that the oceans around the North and South Pole are covered in less ice and for shorter periods of time. This gives impetus to old desires for natural resources that are presumed under the ice, mainly mineral oil and natural gas, which already results in disputes regarding the sovereign rights of the neighboring states, especially in the Arctic. However, the wildlife has adjusted to the special living conditions and can therefore only react to a limited extent to the changes that are introduced by humans and climate change.

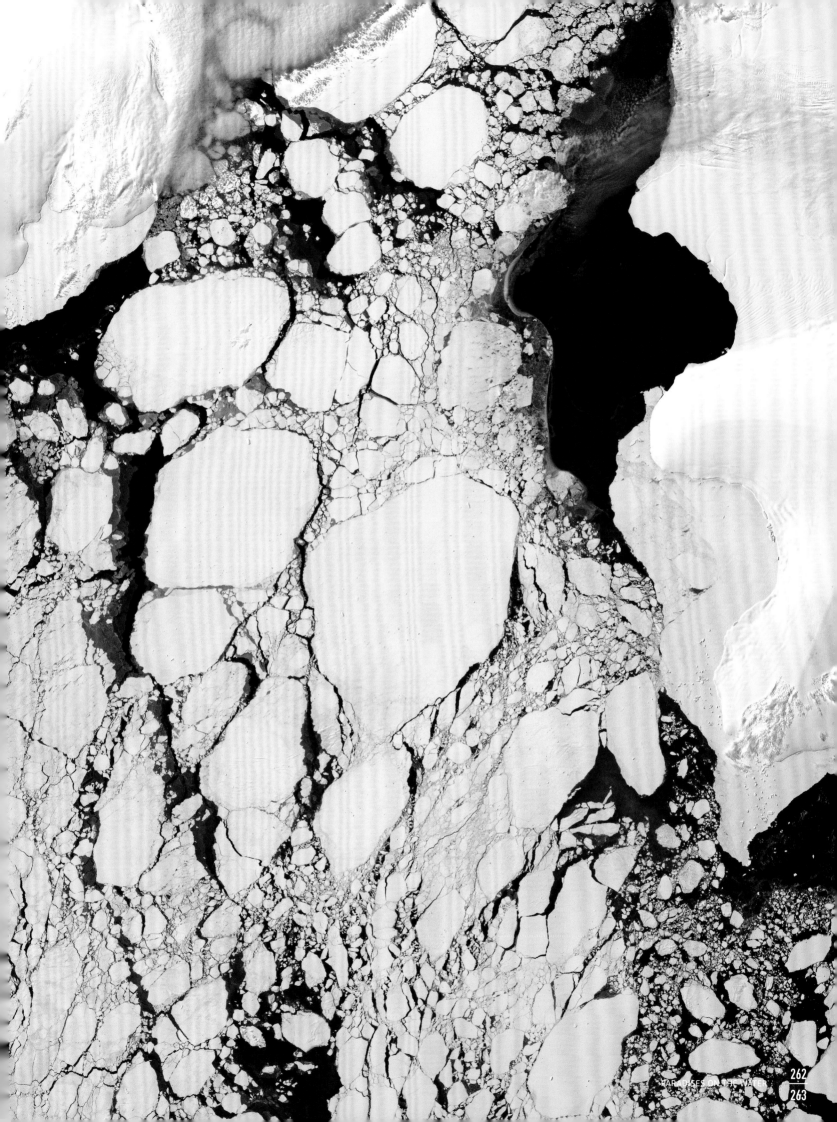

On the Roof of the World

30°46' N · 90°51' E (February 1, 2014)

The salt lake Nam Co is located more than 15,420 feet (4,700 m) above sea level in the middle of the highlands of Tibet. The region had been lifted as a result of the colliding of the Indian and Asian tectonic plates and forms the "Roof of the World" at an average height of 14,760 feet (4,500 m) above sea level. Surrounded by high mountain chains, the highlands are hardly reached by clouds and are therefore very dry.

End of a long journey
45°13' N · 28°44' E (August 27, 2009)

With its area of 2,278 square miles (5,800 sq. km), the river mouth of the Danube at the Black Sea is the second largest river delta in Europe. The world's largest reed bed, an important breeding area for numerous species of birds, is located in the delta. Because of its great importance for biodiversity, the delta was placed under protection as a biosphere reserve.

PARADISES ON THE WATER

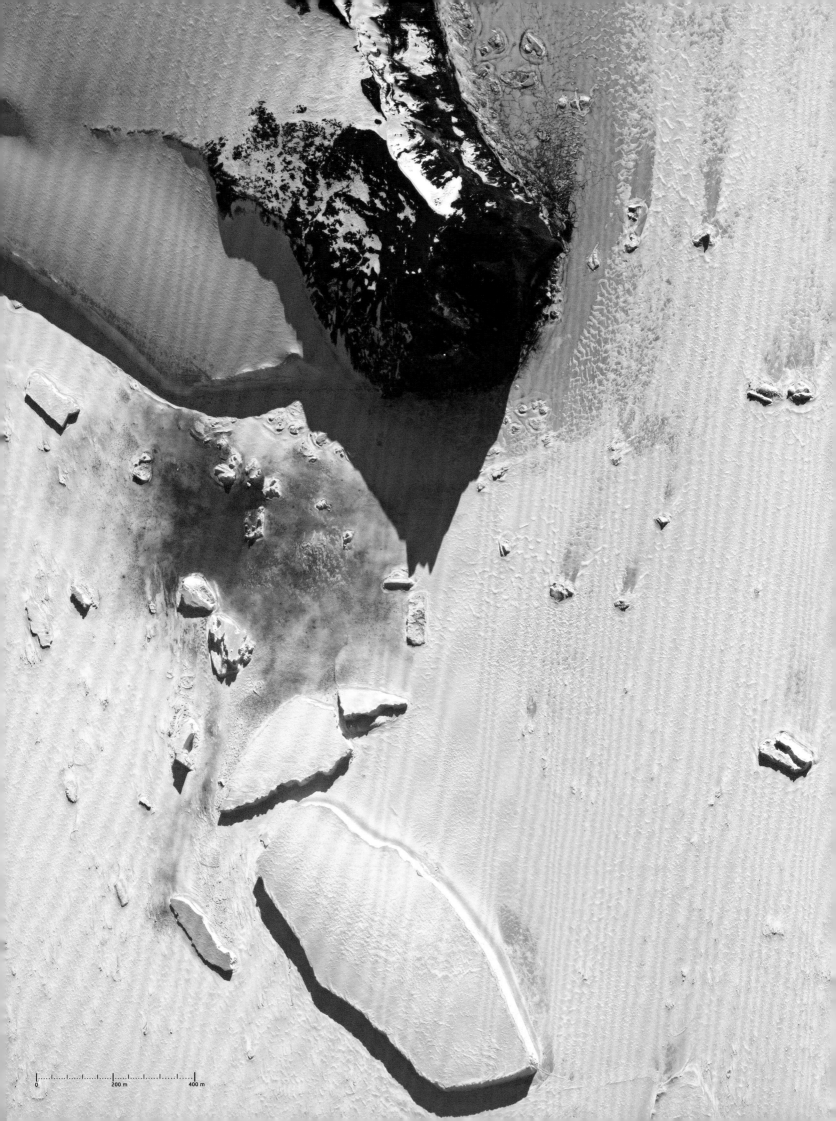

Penguins at the end of the world
74°39' S · 165°25' E (November 8, 2011)

Near the Ross Ice Shelf, the ice surface is soiled by
penguins. Emperor penguins create breeding colonies
with thousands of birds at the edge of the Antarctic.
They are optimally suited for the harsh living conditions and can complete up to 124-mile-long (200 km)
migrations over the sea ice and survive without food
for weeks during their breeding season.

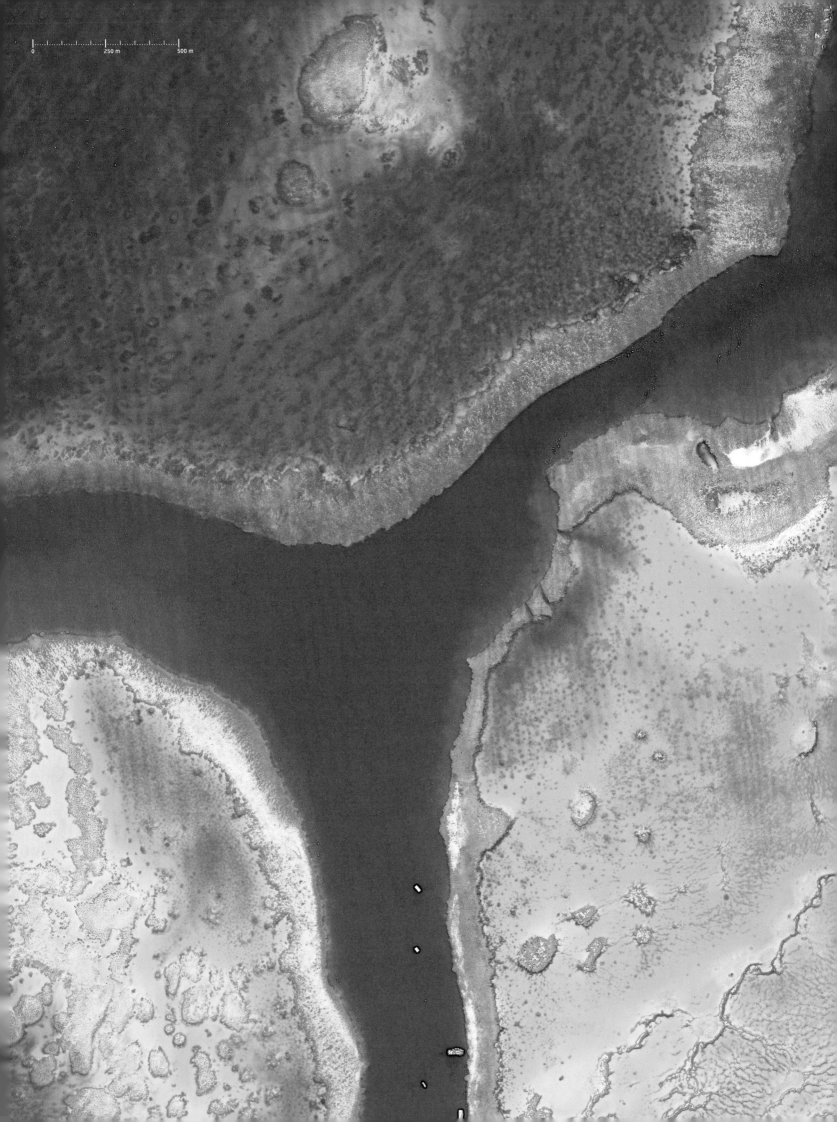

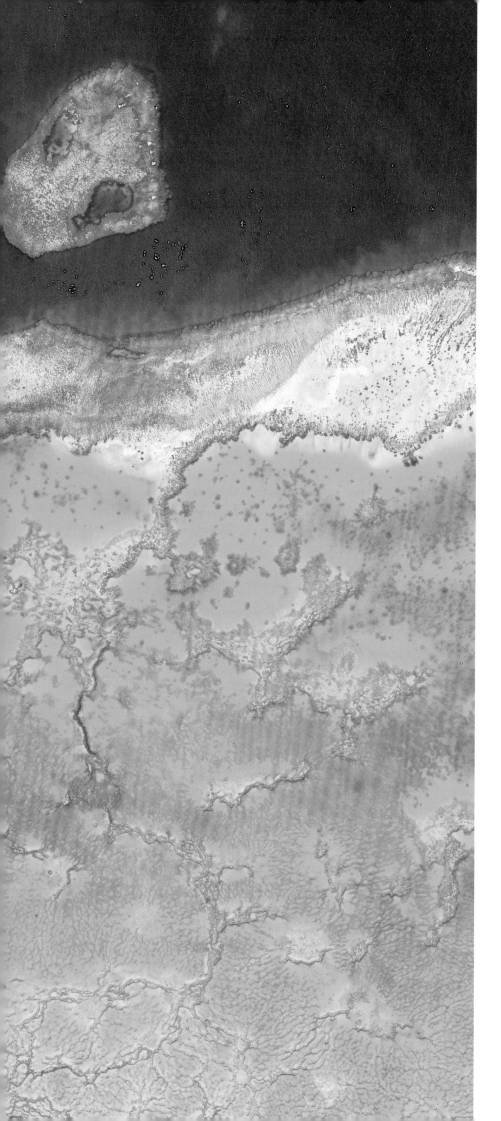

Treasures in the coral reef
19°44' S · 149°11' E (July 29, 2011)

The Hardy Reef, the Hook Reef and the Line Reef
are separated from each other by deep water
areas and form one part of the Great Barrier Reef
off the Australian coast. The shallow water of the
reefs represents an excellent biosphere not for
the corals alone, but also for many mussels, crabs
and species of fish, where they can find food and
shelter from larger predators at the same time.

WHEN CORALS TURN PALE

Corals live in symbiosis with single-cell organisms that
carry out photosynthesis. If the coral is stressed by
unfavorable environmental conditions, it can no longer
provide these single-cell organisms with nutrients and
rejects them. The coral itself then not only loses its color,
but also its energy balance and the coral dies. One of the
factors that can trigger this process is an increased water
temperature. Since the 1980s, there have been repeated
incidences of extended coral bleaching in the Great Barrier
Reef and in partial areas almost complete dying off of the
corals. In addition to coral bleaching, the crown-of-thorns
sea stars that graze the coral reefs in large numbers contribute to the death of corals. The Intergovernmental Panel
on Climate Change (IPCC) fears that the increasing water
temperatures will develop into the greatest threat to the
coral reefs over the next few decades.

THE GARDENS OF POSEIDON
Ocean protection areas for our own protection

Even under water, nature has incredible powers to heal itself. If you want maritime fauna and flora to recover, you simply need to leave parts of the oceans alone — a case for more protection areas in our oceans.

Nature lovers crave to marvel at the wonders of Earth not only in documentaries, but with their own eyes. Anybody who is so lucky to experience the two habitats of our planet with the highest diversity of species — the tropical rain forests and the coral reefs, is stunned and speechless in the case of the coral reefs, while one will often experience some extent of disappointment when visiting the rain forests.

This is because the biodiversity of the tropical rain forest is formed differently to that of the coral reef. The "green hell" offers an unlimited number of species of plants, but only very few specialists among botanists are able to ascribe this glory. And unlike in a documentary, the animal inhabitants hardly show themselves. However, anyone who dives into an only partially intact Indo-Pacific coral reef will still experience thousands of species in the smallest spaces. There is nowhere else on our planet where you can marvel at such diverse, beautiful and such colorful lives.

But other oceans also have their own beauty, even if they cannot match the number of species in the coral reefs. One of these oceans is the European Mediterranean Sea.

Admirers of nature rave about the beauty of the sea. At the same time, they are so naive that they do not realize that this diversity could be lost within two or three generations — destroyed by the changes in nature that are caused by us. For a long time, humans thought that the oceans could absorb an unlimited quantity of waste, waste water and toxins — while we simultaneously and boundlessly exploit its gifts. However, we have long ago reached the limits of resilience of the marine habitats. The Mediterranean Sea, which is a popular vacation paradise, is also one of the most polluted, contaminated and exploited seas of the world.

Let us open our eyes

Before you give way to despair, you should first visit one of the ocean protection areas of the Mediterranean Sea and take a look under water. For example, the small island of Cabrera south of Mallorca which, including the surrounding ocean area was declared a national park in 1991, or the "Parc National de Port-Cros" off the Côte d'Azur, which has existed since 1963.

Under water, observers are presented with a beauty and diversity that has been believed long lost. The ocean floor is covered in a colorful carpet of green, red and brown algae. Thick seaweed areas cover parts of the sandy ground. Sponges amaze with their intense red, brown, yellow, blue and purple. Everything that we find so beautiful in the tropical coral sea also lives here, even if not quite as biodiverse.

Visitors experience large shoals of porgies and barracudas circle in the open water. Every few minutes, a grouper passes by. Colorful Mediterranean rainbow wrasses and ornate wrasses are a reminder of the beauty of tropical sea fauna. Even very experienced divers, who think they have seen it all, emerge from the water after diving here and have a big smile on their faces.

In other areas, the ocean floor is bare, hardly covered in algae, at most you will see a sea urchin and in the best case scenario a small fish passes by.

The order of the day: Don't do anything!

How can you explain this dramatic difference? The answer is simple: In those areas where you cannot find anything anymore, humans have already ruined all life due to destructive fishery methods, waste water and dumped toxins. But the ocean (and nature as a whole) has incredible regenerative powers. What can we do to help it along? Nothing — just leave nature to itself, do not intervene! The reason for this is in a large part that many sea dwellers produce a very high number of eggs (or other forms of reproduction) that turn into larvae, which in turn develop into the descendants as we know them. One single marine animal produces millions, sometimes even hundreds of millions of potential descendants. By no means do they all survive, but that is exactly the real intention — because eggs and larvae of microorganisms, in microscopic small sizes, drift in plankton in unimaginable quantities and provide the basic food resource for other organisms.

When the marine habitat in a protected area is not disturbed, countless millions of eggs and larvae arrive with the plankton and the current, settle, and can develop and produce further. The natural cycles run smoothly, the food webs stabilize and diversity returns. Little by little, other inhabitants arrive. The creatures grow, prosper and send their offspring off on the far journey.

However, this extremely simple but effective regeneration system can obviously only function if there are at least small islands for survival available somewhere, where the species can be preserved. Not even the best ocean protection area can restore a biodiversity that has been lost for good. It has gone past recovery.

Once this is clear to us, then we can better understand the paramount importance of ocean protection areas. The oceans are recklessly overfished and raided, but if several, even small, ocean protection areas provide for replenishment, new ocean dwellers will consistently migrate again to the raided areas with the currents. Studies prove that yields in fishery increase when sufficient protection areas are provided for. The protection areas basically become reservoirs of life, in which biodiversity can be preserved.

The big picture

However, not all problems can be resolved by stopping the utilization of an ocean area and by prohibiting destructive fishery methods. Global factors also have a massive impact on the sea dwellers: climate change, rise of water temperatures, the rising CO_2 content in the atmosphere and oceans and the associated acidification of the water. Additional areas of trouble are the pollution of the oceans with nutrients from agriculture and untreated sewage, the insidious danger from plastic, the invisible death from highly toxic substances that we dump into the environment. Toxins collect in the food webs and finally reach our plates. Eating seafood has become a risky pleasure in many regions of the Mediterranean Sea. Sharks, rays, whales and seals store the toxins in their tissue, which weakens their immune system.

We cannot solve the global problems over night. But in an ocean protection area, you can see results within a short period of time!

Many people still imagine the complexity of the habitat "ocean" and interlinking of things far too simplified. The understanding that we shoot ourselves in the foot when we destroy the biodiversity is missing. Marine scientists no longer refer to food chains — the term "food web" applies more appropriately since it is a highly complex system with an unimaginable number of individual factors. We talk about ecosystems — and systems are something completely different to individual parts. But we are also an individual component of a large system. With every new ocean protection area we achieve something positive for us, our children and grandchildren. In one decade, real miracles can happen in one of those ocean national parks!

Dr. Robert Hofrichter, born in 1957, is a zoologist and biologist, as well as a freelance journalist and nature photographer. Since the end of the 1990s, he has been intensively engaged in protecting the oceans. He is one of the co-founders of the Red Sea Environmental Center, the ocean protection organization mare-mundi and the forum natur & zukunft (which translates to "nature & future"). Hofrichter is associated with the Zoological Institute at the University of Salzburg and author of several books relating to endangered ocean areas.

"We cannot solve the global problems over night. But in an ocean protection area you can see results within a short period of time!"

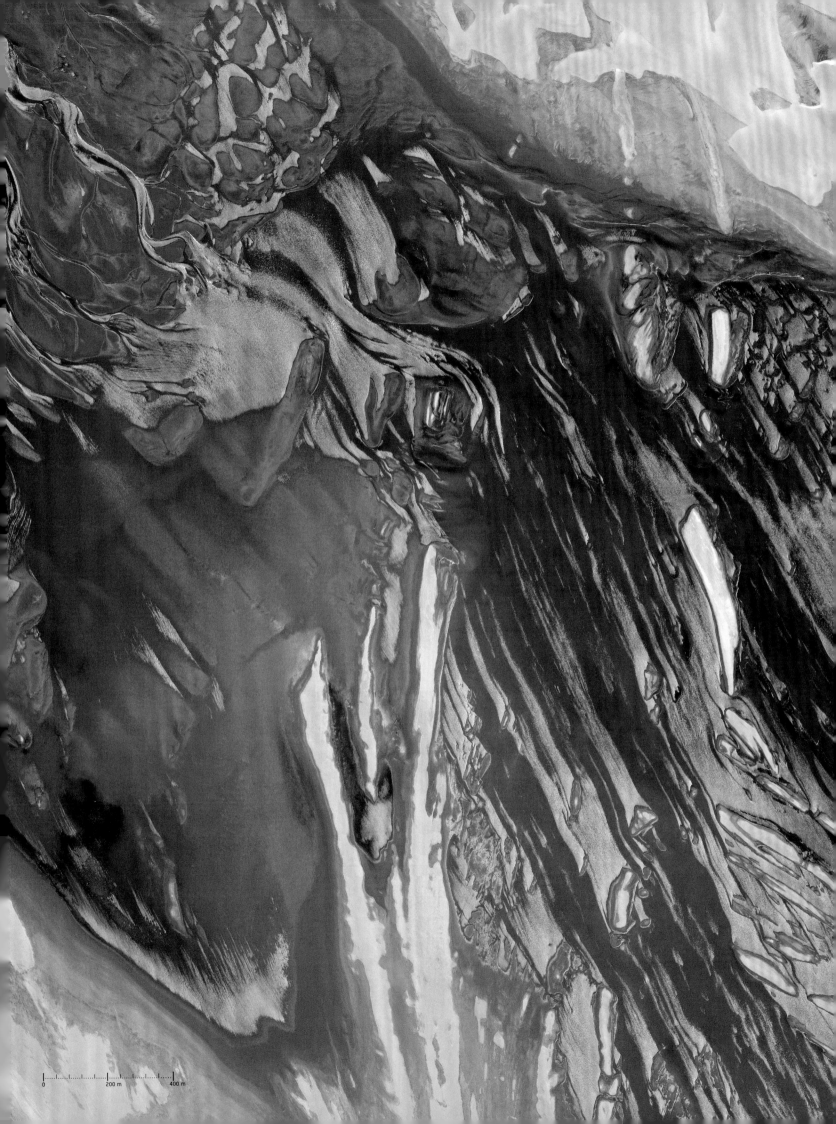

0   200 m   400 m

Bird paradise between the ocean and the desert
23°26' S · 14°28' E (February 17, 2006)

The bay of Sandwich Harbour is located south of
the Namibian whale fishing bay between the sand
dunes of the Namib Desert and the Atlantic Ocean.
Sandwich Harbour is developing more and more into a
lagoon because of sanding-up. Large bird colonies live
in the flat bay, including flamingos, terns, pelicans and
cormorants.

Colorful jewels
44°32' N · 110°50' W (July 13, 2010)

Parts of the Yellowstone National Park, which was created in 1872 as the first national park worldwide, are
located above the magma chamber of Yellowstone's
super volcano. For this reason, you can find many hot
springs and geysers here. With a diameter of 295 feet
(90 m), Grand Prismatic Spring is one of the largest
thermal springs in the world. It owes its play of colors
to the bacteria and algae that live in the respective
temperature zones suited for them.

Between rough and calm waters
8°56' S · 158°02' W (January 14, 2010)

The lagoon of Tongareva is well protected against the
rough sea. Only small boats are able to cross the flat
Siki Rangi Passage in the calm waters of the lagoon.
The atoll protrudes almost 16,400 feet (5,000 m) from
the ocean floor, yet is only elevated a few feet above
sea level. The previous vegetation of the island has
mostly been replaced by coconut palms — the basis
for copra production, one of the few export products
of the island.

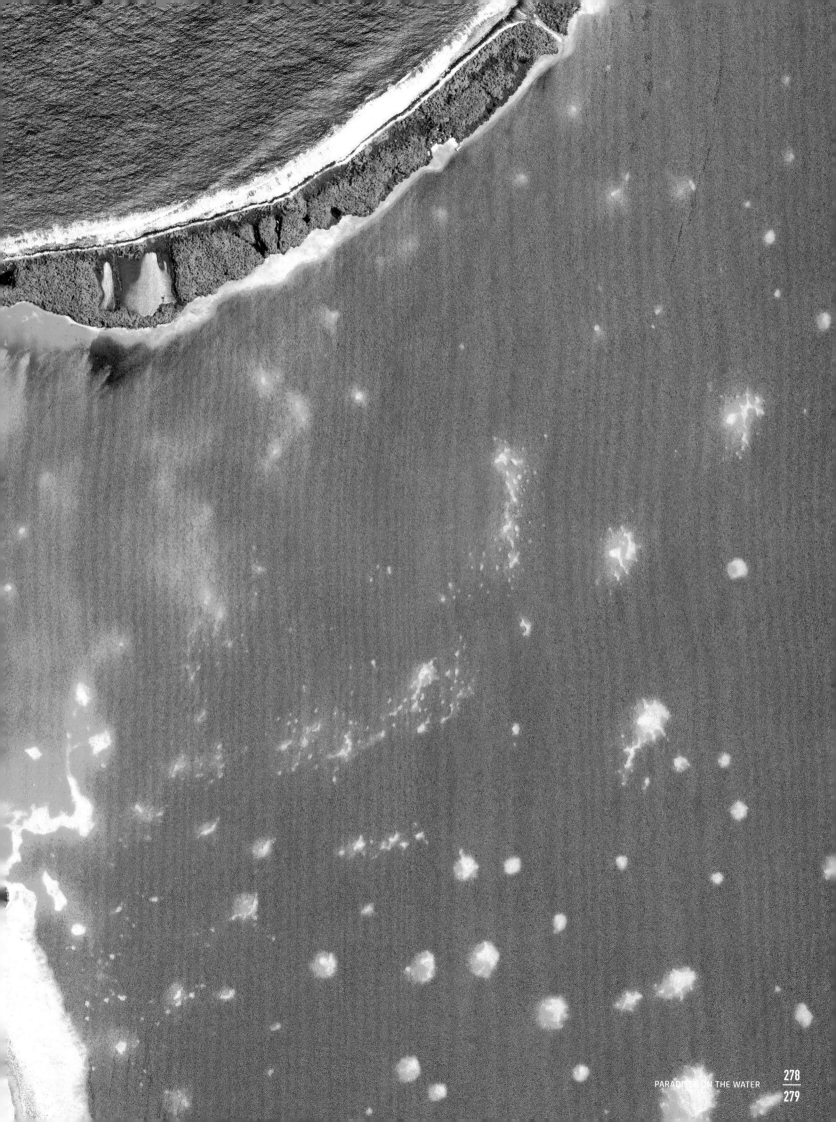

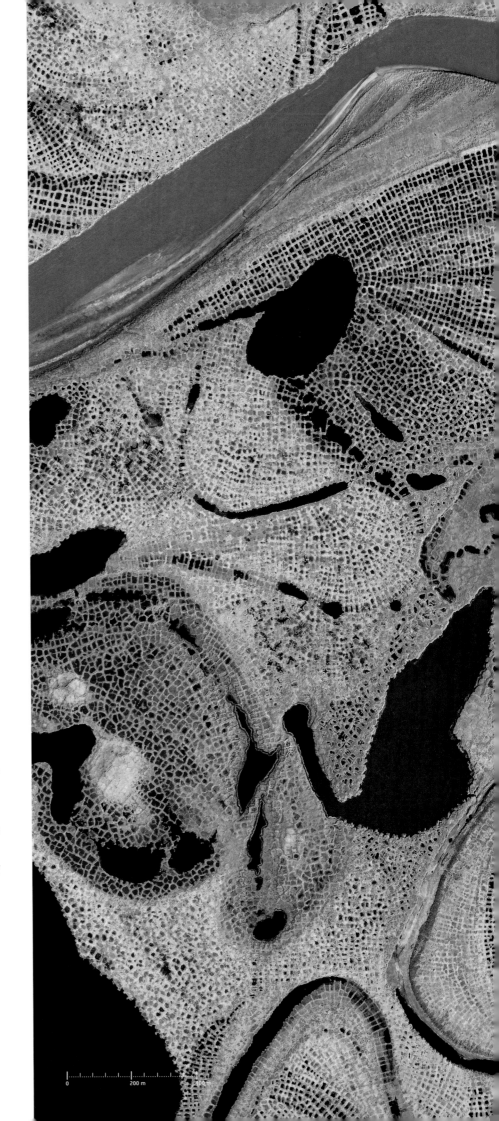

Large areas of Eastern Siberia are characterized by
permafrost; this also applies to the delta of the
Yana River, which flows into the Yana Bay here.
The extent of the frost's impact in this area is also
demonstrated by the fact that the Yana Bay is
covered in ice for up to nine months per year. The
netlike structures of the ground and the numerous
lakes are created as a result of repeated freezing
and thawing of the tundra.

HEAT-TRAPPING GAS IN
PERMAFROST

Almost one-quarter of the land surface of Earth is covered
by permafrost, especially in the regions around the Arctic
Ocean. In all climate models, permafrost grounds play
an important role since an increase of temperatures may
result in thawing of the frozen ground. On one hand, this
would result in a loss of the stability of the ground, thereby
resulting in buildings becoming unstable and landslides in
mountain areas. At the same time, large quantities of carbon dioxide and methane that are currently bound into the
biomass of the ground would be released. These are gases
that contribute to the warming of the planet in the form of
greenhouse gases. Methane even has 25 to 33 times the
global warming potential when compared to carbon dioxide. Signs for the beginning of permafrost ground thawing
can be observed in Siberia and Alaska.

0   200 m   400 m

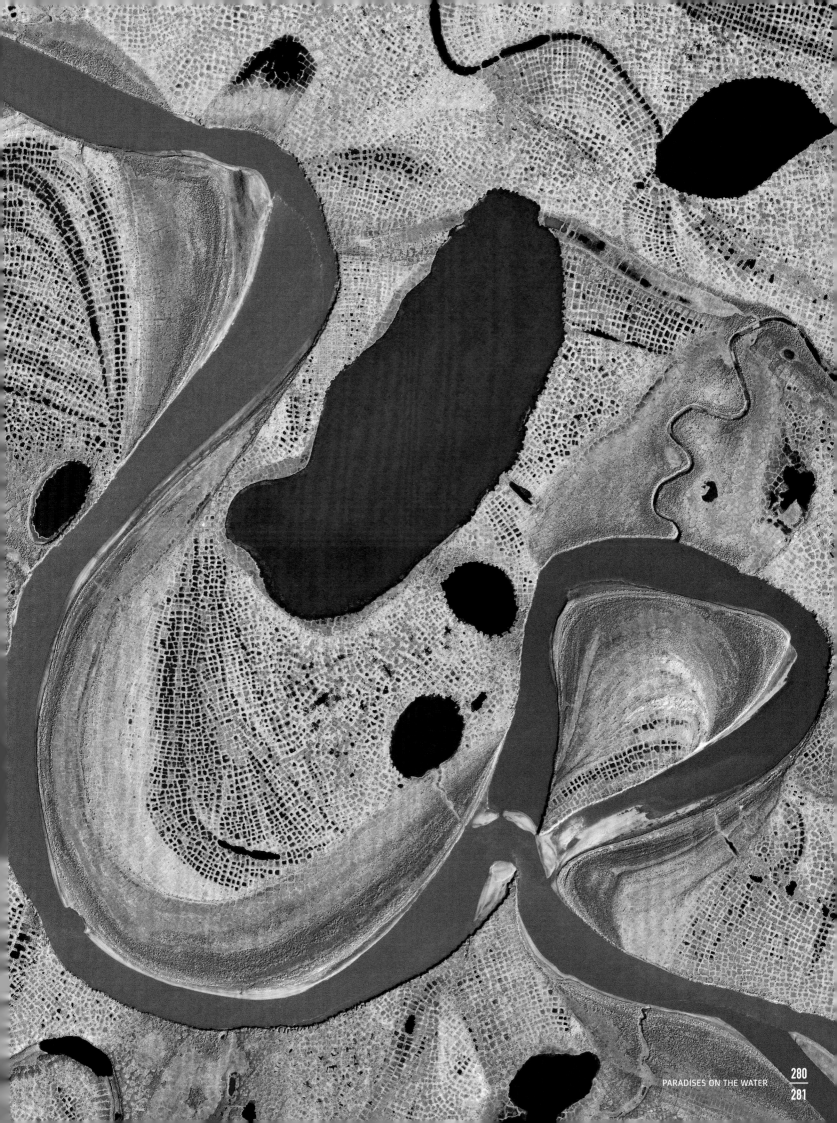

Laboratory of evolution

0°17' S · 91°21' W (August 17, 2009)

The volcanic Galapagos Islands are located approximately 620 miles (1,000 km) off the coast of Ecuador. Their isolated location resulted in the development of a very individual flora and fauna, which was already used by Charles Darwin to support his evolution theory. Known examples include giant tortoises and marine iguanas that dive up to 30 feet (9 m) deep into the ocean when searching for algae.

Fire, water and ice
64°39' N · 18°38' W (June 23, 2013)
The Múlajökull is an extension of the third largest glaciers of Iceland, the Hofsjökull, which covers the largest active volcano in the center of the island. As shown by the concentric moraines around the glacier edge with their numerous small lakes, this glacier is retreating. Here, the moor area Thjórsáver with the largest breeding colony of the pink-footed goose adjoins.

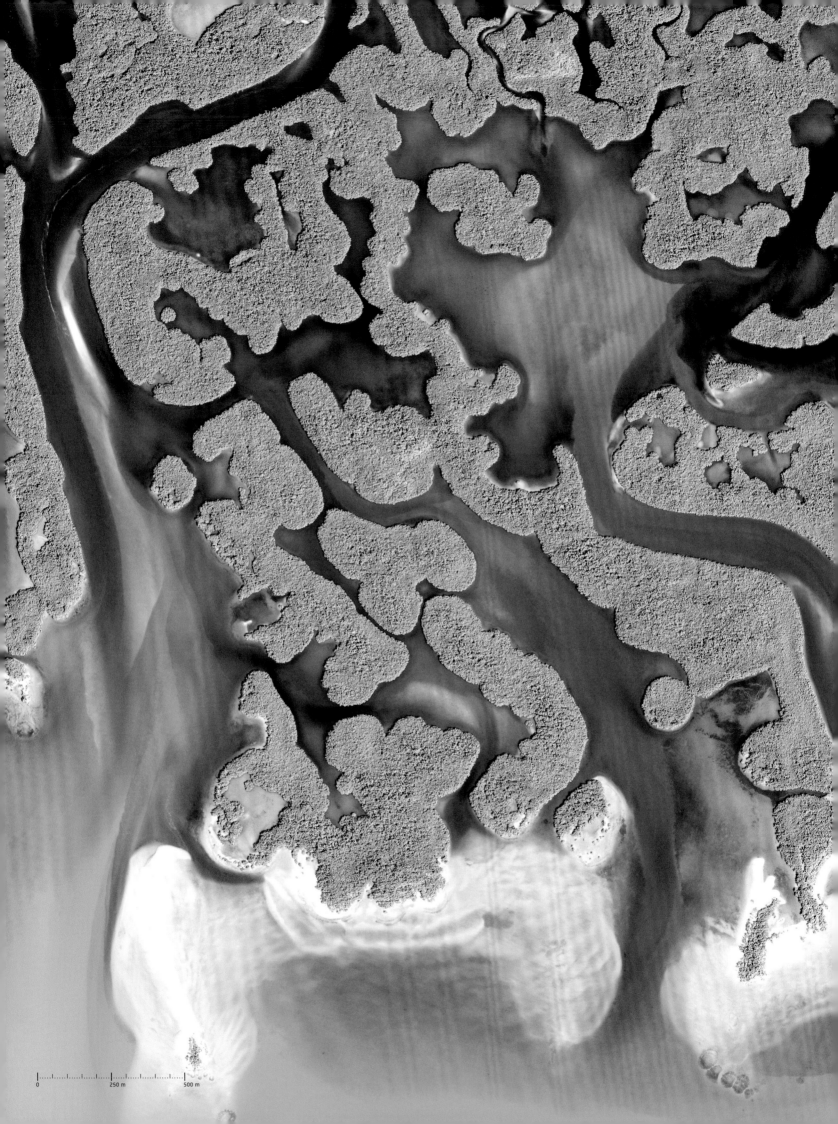

Home to crocodiles and alligators
25°51' N · 81°30' W (January 11, 2010)

The flat swamps of the Everglades in Florida are mainly covered in grass areas that are adjoined by mangrove forests at the coast of the Gulf of Mexico. The mangroves form a protection belt against the erosion forces of the sea during the frequently occurring hurricanes. This coastal region in Florida is also the only area where you will find both alligators and crocodiles.

MANGROVES — PROTECTING NATURE

Along the shore areas of tropical and subtropical oceans, you will often find mangroves; these are forests that are composed of trees with a salt tolerance. These trees are able to use the salty water at the edge of an ocean, by either not even absorbing the salt or by releasing it again. As a result of this, they are also able to handle alternating water levels that are changing due to ebb and flow. By means of stilt roots, mangroves are able to absorb oxygen directly from the atmosphere, which is then transmitted in the underground roots. The trees fulfill several functions: Besides the protection of coastal areas against the sea surf, they provide ideal breeding conditions for many species of fish, crabs and mussels because of their dense and flooded network of roots. Since they are not directly usable, mangroves are endangered especially because of the expansion of aquaculture areas for shrimp farming.

PARADISE ON CALL
A large wetland area is endangered in Brazil

Birds, reptiles, insects: The Pantanal in southwestern Brazil is a marshland with a huge diversity of species. But the insidious contamination with mercury could mean the end of the tropical wilderness.

One of the largest wetlands in the world is located in the heart of South America. It is called Pantanal, which means "flood area." Its size is almost equal to the size of West Germany before its reunification. Only the western Sibirian lowlands are more extensive, but they are frozen during most of the year. The third competitor for the title of "Largest Wetland on Earth" is the area of the swamps at the upper Nile in South Sudan, the infamous Sudd.

The Pantanal is located between the central Brazilian highlands and the southeastern Bolivian hills at the upper Paraguay River. To the south, it borders on the edge regions of the Paraguayan Grand Chaco. It is considered the last wilderness of South America and the most interesting one with respect to its fauna. Here, you can experience wild animals better than at the best spots of Amazonia – despite the fact that a large part of the area consists of swamp and pastures and has less of a "wilderness" atmosphere than the Amazonian rain forests that have not yet been cut down.

Giant guinea pigs and other impressive animals

Taking a trip on the Transpantaneira highway will give you the first impression. You should avoid the rainy season, because then, vehicles are at risk of sinking into the mud and most of the 122 bridges along the 90-mile (145 km) route, which are quite ailing anyway, are flooded. During the dry season however, the Transpantaneira is breathtaking for South American conditions and only comparable to Africa. Dozens, and sometimes hundreds, of caimans are lying at the trenches directly next to the road and only reluctantly move into the water when a vehicle appears. Startled and disturbed in their siesta at the edge of the road, hoards of ragged brown animals quickly run into the swamps. If you get the impression that these animals look like giant guinea pigs then you are not wrong. The capybaras, as they are called in Brazilian, are actually relatives of the guinea pigs. In German, the rodents are called "water pigs." At up to 110 pounds (50 kg), they exceed our largest beavers by almost twice the weight. You might also spot a giant snake, 16 to 20 feet long (5 to 6 m), that is sunbathing on the road. The anacondas, which are not rare in the Pantanal, like to do that because the heat helps their digestion if they have just swallowed a whole capybara.

Wild bird views

You can also observe many birds along the track. The southern lapwings are very vocal birds. They have thornlike, bony extensions (spurs) under the wings. Sometimes, they use these spurs to stab a cow in the mouth if it accidentally got too close to their nest of eggs or their hatchlings. You will almost always see groups of loud screeching parakeets or Amazon parrots that usually fly by in pairs, and ducks, herons, storks, ibises and other aquatic birds, gather around the larger lagoons by tens of thousands. In the Pantanal, ornithologists can observe the objects of their desire with a good pair of binoculars far better than in the forests of Amazonia. Although the diversity of species is considerably lower, 665 different ones were still counted and observing them is clearly easier.

About hunters and the hunted

You will, however, rarely see the most attractive animal of the Pantanal: the jaguar. The Spanish-speaking neighbors of Brazil call it *El Tigre*, full of awe. In the Portuguese-speaking Brazil, it has the Brazilian Indian name *Onça pintada*, which means "spotted Onça." Pumas also roam through the underbrush of the Pantanal. Both big cats are considered a danger to the cattle, which mostly grazes freely on large *facendas* (cattle ranches). Some of these farms are the size of an entire German province; individual ones are even larger. The jaguars and pumas consider the cattle as a welcome part of their menu. No wonder, because before the Spanish and Portuguese introduced the first cattle, no comparable large animals existed in South America. The capybaras are some of the largest mammals; there are only a few other long-established species that are larger: for example the two South American tapir species, the swamp deer and brockets, as well as the guanacos — small camels related to lamas that live further away in the south in the pampa. Biologists have not yet been able to understand why the South American mammals are all considerably smaller than their counterparts in Africa or Southeast Asia. The South American nature is actually quite suited for large animals. You can observe that by how easy it was to naturalize cattle, horses and sheep, as well as European game in South America.

In Amazonia, you will find a particularly low number of large animals. This is due to the natural conditions. The soil is so meager that they cannot sustain any permanent cultivation. Once a forest has been cut down, the new pasture areas are neither suited for long-term grazing by cattle nor for growing soy, which is used as power food for our European indoor livestock. This is different in Pantanal. The rivers that originate from the Brazilian highlands keep the Pantanal fertile. For centuries, pastoral farming has been successfully practiced. And although jaguars, who quickly found cattle more attractive than capybaras, were hunted vigorously from the beginning, the big cat population grew rather than shrank. Cattle became abundant prey. In the Pantanal, "tiger safaris" have also been offered for a long time. But instead of rich big-game hunters, who pay a lot of money for one jaguar kill, more and more nature tourists are visiting the wilderness, and that is more profitable in the long run. There are, however, sufficient cattle in the Pantanal and the annual floods and subsequent very hot dry seasons set limits to cattle ranching.

Treacherous threat from the mountains
So, does that mean the Pantanal is in a good position? Unfortunately, no. For the main part, neither big game hunting, which is also socially increasingly outlawed in Brazil, nor the increase of cattle farming, affects the huge wetlands. And the growing number of tourists is also not a problem. They visit to experience and take pictures of magnificent birds, such as the huge hyacinth macaw in the wild. On the contrary: Because the visitors spend a lot of money for meeting the large blue macaws, some farmers now protect the largest species of parrots. No, the actual threat of the Pantanal drifts mostly unnoticed with the water down the rivers from the mountains from Mato Grosso. It spreads like a slow poison, accumulates, takes effect and kills. It is actual poison, one of the most persistent and malignant of all: mercury. This is because, at the top, in the mountains of Mato Grosso, gold is washed from the sand of the rivers and during this process; the toxic metal is used abundantly and hardly controlled. Torrential rains transport it into the Pantanal. However, mercury does not decompose. It remains highly toxic.

So the gold with which the prosperous treat themselves carries the slow death into the Pantanal. To the same extent as previously reliable currencies are devalued by speculations or debt overloads of states, the value of gold increases. The (value) consistency of gold is equal to the (poison) consistency of mercury. Consequently, the future of the Pantanal, this unique irreplaceable tropical wetland area, is determined, to a greater degree, by those of us living outside of Brazil.

Professor Dr. Josef H. Reichholf, born in 1945, is a zoologist, evolution biologist and ecologist. He has been working at the Bavarian State Collection of Zoology ("Zoologische Staatssammlung") in Munich for over 35 years. In addition, he taught at the University of Technology and Ludwig Maximilian University of Munich. Reichholf is member of the commission for ecology at the Bavarian Academy of Sciences and Humanities ("Bayerische Akademie der Wissenschaften") and is author of numerous scientific and popular scientific publications. He is one of the most important natural scientists of Germany.

"Instead of rich big-game hunters who pay a lot of money for one jaguar kill, more and more nature tourists are visiting the wilderness."

Wet nature paradise
17°37' S · 57°41' W (July 14, 2007)

With an area of approximately 77,000 square miles (200,000 sq. km), the Pantanal, which is mainly located in Brazil, is one of the largest wetland areas of the world. During the rainy season, water drains only slowly from the flat level land so that 80 percent of the area can be up to 16 feet (5 m) under water. The fauna and flora of the Pantanal is protected in nature reserves, however, agricultural use of the region poses a threat.

0   200 m   400 m

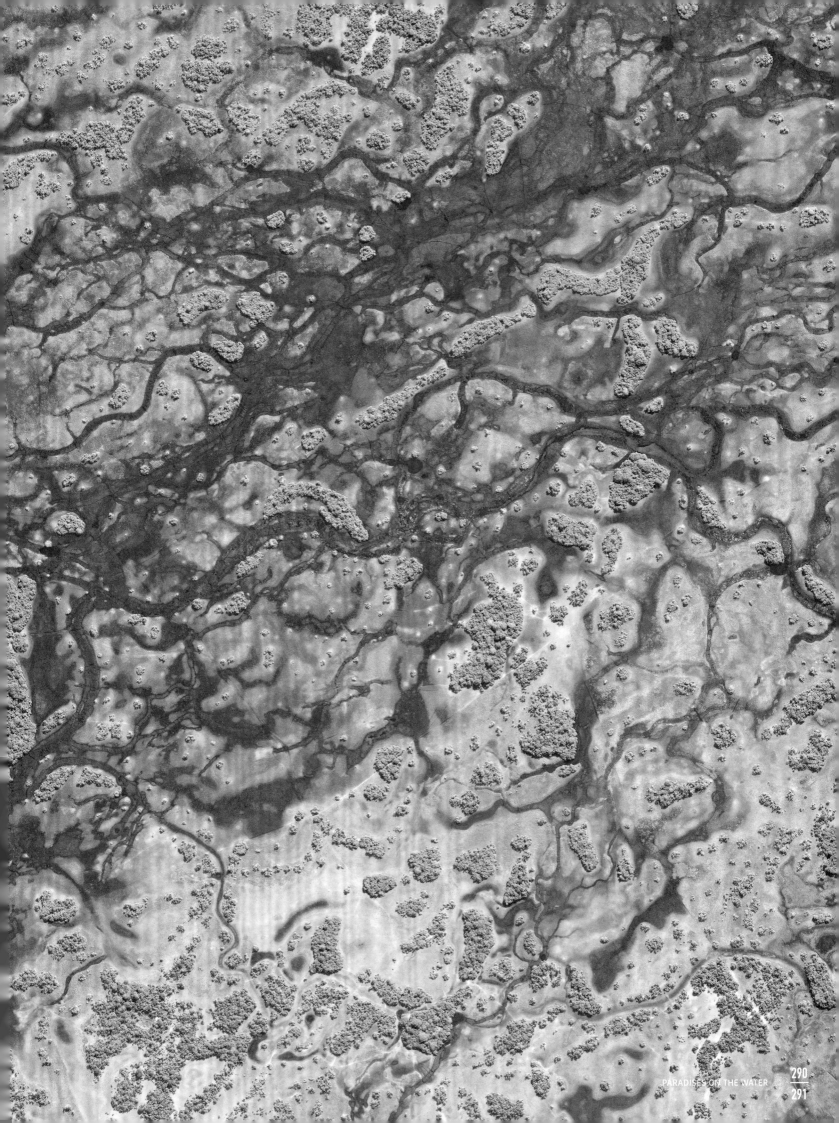

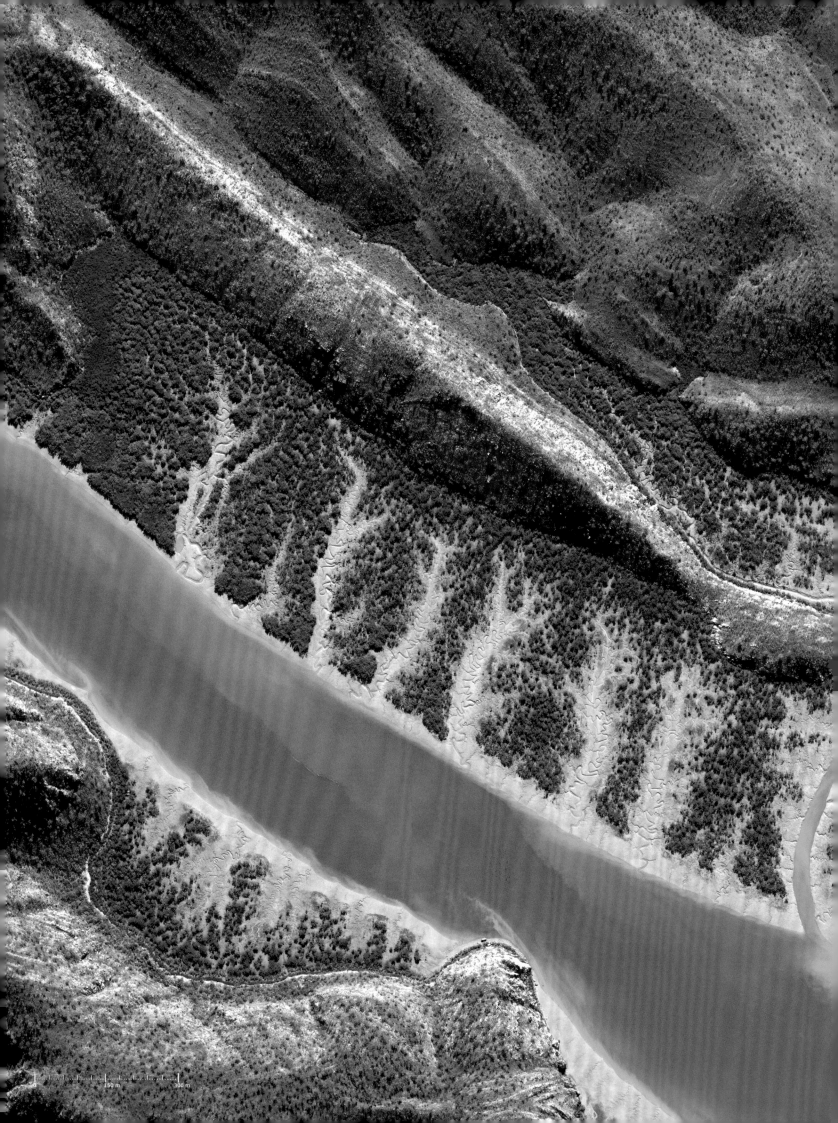

Mangroves between ebb and flow
16°13' S · 123°47' E (July 7, 2011)

Round polished rock formations shape the coastal landscape of the northwestern Australian Kimberley with its many islands off the coast. The bays are clearly formed by the turning of ebb and flow that are very strong here and provide habitat for many species of animals with their mangrove forests.

The Glaciar Upsala stretches from the Patagonian Cordilleras over 37 miles (60 km) to the Lago Argentino and is one of the largest alpine glaciers with a dimension of 218 square miles (565 sq. km). Icebergs calve from the almost 200-foot (60 m) high and 2.5-mile (4 km) wide rim at the glacier end, and float across the lake. The surface and color indicate the dynamics of the glacier that is decreasing in size like many others.

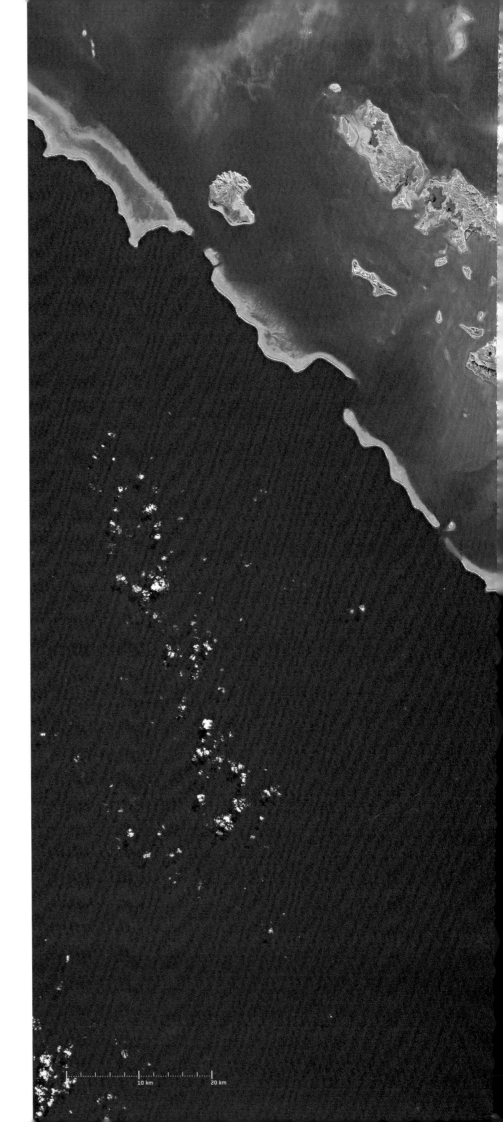

Where the manatees graze

20°11' S · 164°16' E (May 10, 2001)

Grand Terre, the main island of New Caledonia is surrounded by an approximately 932-mile (1,500 km) long barrier reef. Especially in the northwest, where the reef pushes over 125 miles (200 km) into the sea, it creates an extended lagoon that is home to dugongs, turtles and other species of animals. Due to its diversity of species, the lagoon has been placed on the World Heritage List by the UNESCO.

EVOLUTION IN THE DEPTHS OF THE OCEAN

The deep sea that is hard to reach even with technical aid may still be explored with considerable efforts. Over a long period of time, it was considered a scarcely animated zone on our planet, where, due to the high pressure and lack of light, only very few, highly specialized species of animals find their habitat, the bizarre looks of which were used as inspirations for fantasies of deep-sea monsters. Over the past few decades, this view has changed considerably. With the "black smokers," volcanic hot water vents along the ocean ridges, areas at the ocean floor were detected where life exists even at temperatures of over 752°F (400 °C). Now, the "black smokers" are at the center of interest as a special laboratory of evolution. It is possible, that it was exactly those "black smokers" that provided the conditions that got life on Earth started.

10 km   20 km

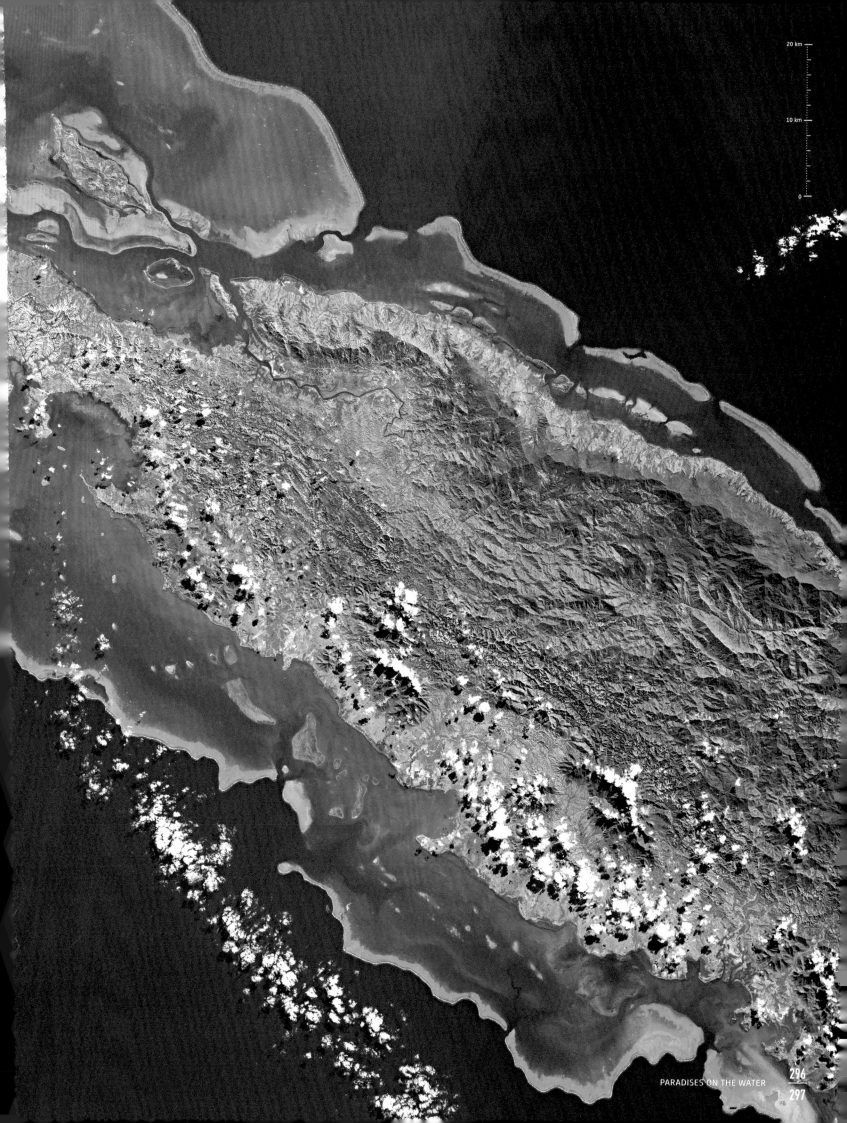

20 km

10 km

0

Variable landscape

26°15' S · 122°39' E (May 7, 2003)

Lake Carnegie is located in the middle of West Australia in a closed hollow. The area is covered by muddy marsh most of the time and the lake fills with water only during periods of significant rainfall, where the light color clearly shows how shallow the water is. During these time periods, the desert around it blooms for a short time.

0   5 km   10 km

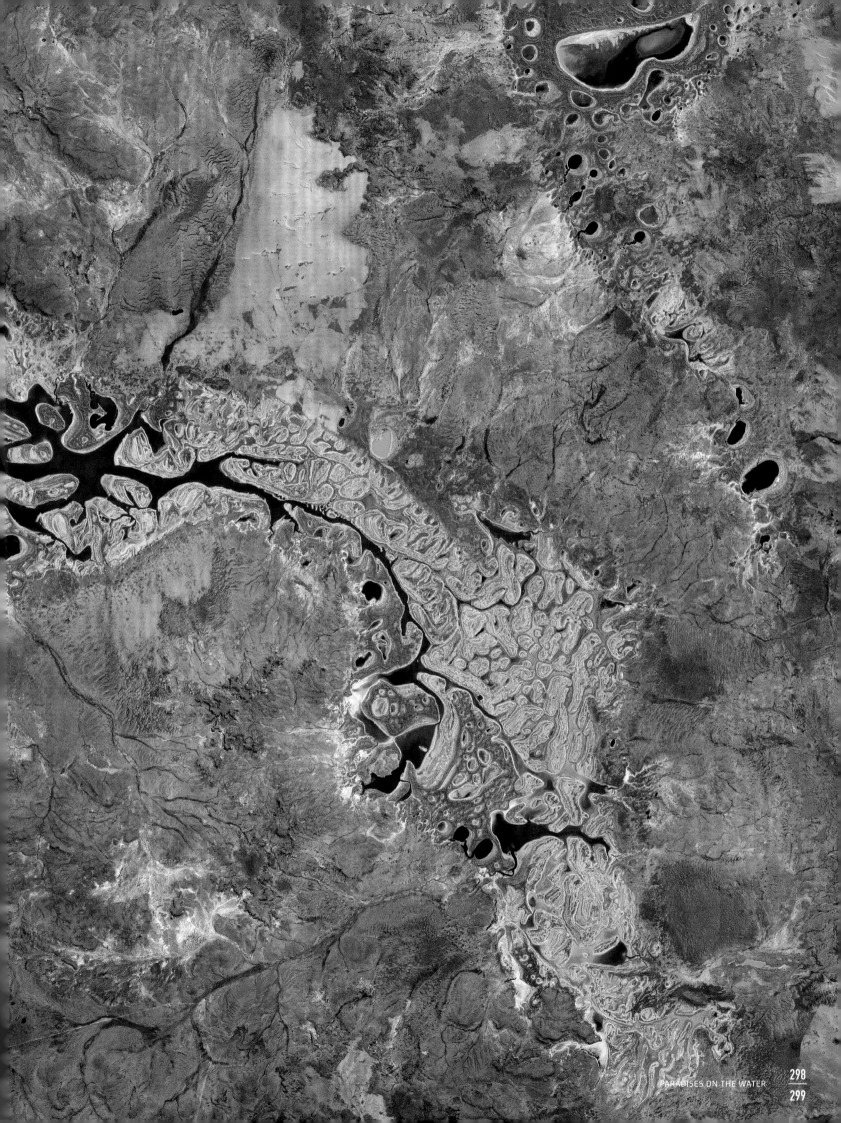

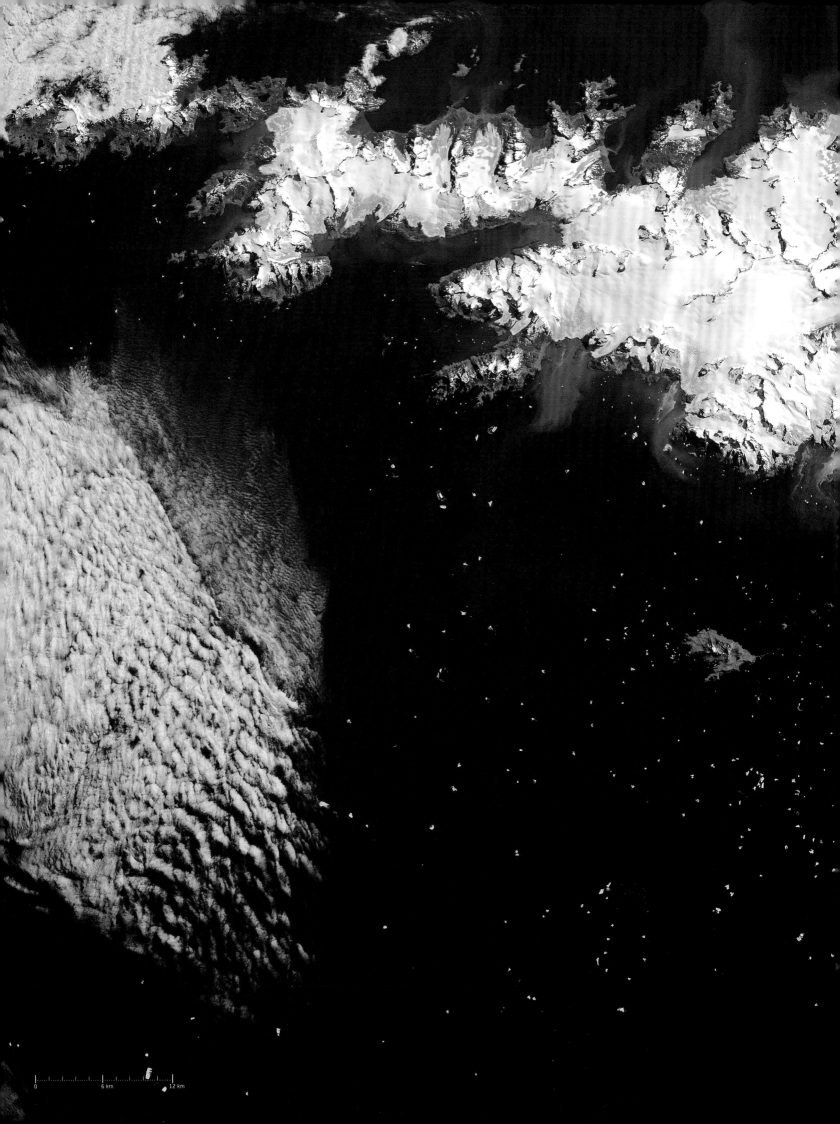

0   6 km   12 km

Glacier island in the stormy sea
54°25' S · 36°52' W (February 07, 2003)

The mountainous island of South Georgia, which is covered by glaciers over large parts, is located in the South Atlantic approximately on the same latitude as Tierra del Fuego. Millions of macaroni penguins live on the island and you will also find the most important breeding area for the king penguins here. Different species of seals, including elephant seals and fur seals, defy the inhospitable climate of the island, where only a few people live.

How an illustrated book with satellite images is made

SNAPSHOTS FROM SPACE

By means of satellite images, it is possible to present even remote areas of the world that can only be reached by photographers with great difficulty, and which can be presented in a consistent form. The quality of the high-performance cameras on board Earth observation satellites, the efficient software for processing the recorded data (and the profound knowledge about the processing of this data) and the stories that are hidden behind these images are of vital importance.

On the way from the idea for an illustrated book with satellite images to its realization, several obstacles have to be overcome. That is because satellite images are only marginally made with the intention of producing esthetically appealing images of regions that are otherwise, more than likely not at the center of interest. The intention behind producing satellite images is usually more related to illustrating conflict areas in order to master humanitarian disasters, determining the state of areas under cultivation and nature protection areas, creating planning bases for infrastructure projects, or monitoring shipping routes.

High-tech cameras in space

For the most part, Earth observation satellites are high-quality cameras that circle in near-polar orbits around Earth and take images with partially extreme telephoto lenses. Due to their productive efficiency, these telephoto lenses are actually more like telescopes. They are needed because satellites take images from distances of 250 to 560 miles (400 to 900 km) in altitude. Here, they are basically no longer exposed to the breaking force of the atmosphere and can operate for years.

In order to illustrate camera properties, we would like to provide you with the data of a modern image-taking system: Satellite WorldView-2, the data of which was used, among others, for this book, has a focal length of 43.6 feet (13.3 m) and the aperture diameter is 3.6 feet (1.1 m). This is sufficient to resolve details of 2 feet (0.5 m) from a distance of 422 miles (680 km). For this purpose, the satellite uses a sensor that records a width of over 35,000 pixels, which would correspond to approximately 900 megapixels per image with a common digital camera, depending on the picture format.

Who releases the shutter?

So when we talk about satellites being high-tech cameras in space, then obviously two questions arise: Who determines the motive? And who releases the shutter? For a long time, it was very time-consuming and complicated for customers of satellite operators who were interested in data to obtain images of a requested area.

For a long time, there were only very few satellites that were even able to produce images in the requested quality. Satellites that are able to display details of a few feet or even smaller, have only existed for 15 years. In addition, these systems were not designed for being customer-friendly. Anyone who ordered current images of a certain region often had to wait for the results for months. And when you then finally received the requested data at a very high price, it frequently occurred that the area of interest was covered by clouds.

This situation has significantly improved with today's developed satellite systems. It may still take months until you receive a usable, up-to-date image of a requested region, but meanwhile, extensive data archives exist in which you can often find suitable images for your specific purposes. Using images that were specifically taken for the task of this illustrated book would have been far too expensive and consequently, we resorted to using images from archives.

So while the search for a suitable motive is pursued similarly as in a classic photo research undertaking, in our case, the release of the shutter corresponds more to searching for suitable images in the archives. The selection criteria are quite similar to those of classic photography: The content of the image must match just as well as the lighting conditions, that is, the effect of the different light and shadow conditions depending on the season.

From a data record to a satellite image

Similar to classic photography, the production of satellite images requires some processing steps that, however, deviate considerably. Modern Earth observation satellites usually do not transmit color images, but a series of data records or image channels taken in different wavelength ranges. By means of image processing and special software for processing satellite data, these data records are converted to images for which, in addition to the color composition that is accomplished by appropriately mixing the available image channels, correcting the geometry of the image also plays a role. Similar to a map, the corresponding excerpt should display the surface of Earth with as little distortion as possible.

In further optimization steps, the influence of mist on color and contrast is minimized and the best possible image sharpness striven for. This can often be achieved by combining a black-and-white image that has been taken with a higher detail sharpness, with color information of a lower detail sharpness. The goal of all these processing steps is to obtain a true-to-life rendition in brilliant colors that equals viewing it at close range.

From a satellite image to an illustrated book

After developing the idea for such an illustrated book with satellite images, a rough conceptual design must first be developed; this will define the main framework with respect to topics and content. This conceptual design is followed by extensive content research during which interesting topics are collected — always with respect to the possibility of displaying these topics in the form of satellite images. During comprehensive searches of the available archives, the best suited data records are selected, which are then processed to images as described above. Before inserting the images into the previously defined layout of the book, their respective optimal angle and image section is determined. During the last production step — printing the book — obtaining the best possible color quality takes center stage.

This color quality is therefore subject to special controls.

The vision on which this book is based is fulfilled once the mixture of compiled satellite images covers an appealing spectrum from the point of view of content, as well as esthetics.

↓ **Taking a sharp image of Earth**

QuickBird is part of a group of earth observation satellites that provide images with a high accuracy of detail. They reproduce details of 1 to 3 feet (30 cm to 1 m) in diameter and can therefore be used for detailed surveys of conditions. The photomontage of the satellite above the eye of Hurricane Katrina demonstrates the option of using satellite data for disaster operations.

PICTURE CREDITS

INDEX